T0320340

Wastewater Treatment with the Fenton Process

The presence of refractory organic compounds in wastewater is a global problem. Advanced oxidation processes, in general, and the Fenton oxidation process are alternative technologies for wastewater and water treatment. This book gives an overview of Fenton process principles, explains the main factors influencing this technology, includes applications, kinetic and thermodynamic calculations and presents a strong overview on the heterogeneous catalytic approach. It demonstrates that the iron-based heterogeneous Fenton process, including nanoparticles, a new complex solution, is highly efficient, environmentally friendly and can be suitable for wastewater treatment and industrial wastewater.

FEATURES

- Describes in detail the heterogeneous Fenton process and process applications
- Analyzes the advantages and disadvantages of different catalysts available and their suitability to specific processes
- Provides economic analysis of the Fenton process in a ready-to-use package for industrial practitioners for adaptation into already existing industrially viable technologies
- Promotes a modern solution to the problem of degradation of hazardous compounds through ecological and environmentally friendly processes and the use of a catalyst that can be recycled
- Explains highly complex data in an understandable and reader-friendly way

Intended for professionals, researchers, upper-level undergraduate and graduate students in environmental engineering, materials science, chemistry, and those who work in wastewater management.

Dominika Bury is a PhD student at the Warsaw University of Technology (WUT), Faculty of Materials Science and Engineering, Department of Ceramic and Polymer Materials. Her research focuses on the efficiency of the heterogeneous Fenton process with iron catalysts in degradation cosmetic wastewater.

Michał Jakubczak is a PhD student at the Warsaw University of Technology (WUT), Faculty of Materials Science and Engineering, Department of Ceramic and Polymer Materials. His research focuses on ecotoxicology and antibacterial properties of 2D nanomaterials, mostly MXenes.

Jan Bogacki, PhD, is an assistant professor in the Department of Informatics and Environmental Quality Research, Faculty of Building Services, Hydro, and Environmental Engineering at the Warsaw University of Technology (WUT). He earned a PhD at the Warsaw University of Technology and is the author of 50 scientific publications.

Piotr Marcinowski, PhD, is an assistant professor in the Department of Informatics and Environmental Quality Research, Faculty of Building Services, Hydro, and Environmental Engineering at the Warsaw University of Technology (WUT). He earned a PhD in 2002 at the Warsaw University of Technology and is the author of 40 scientific publications.

Agnieszka Jastrzębska, PhD, DSc, is a professor at the Warsaw University of Technology (WUT). She leads an interdisciplinary research team at the Faculty of Materials Science and Engineering, Department of Ceramic and Polymer Materials. Her research concentrates on nanotechnology for bioactive materials with multi-functional hybrid nanocomposite structures.

Emerging Materials and Technologies

Series Editor: Boris I. Kharissov

The *Emerging Materials and Technologies* series is devoted to highlighting publications centered on emerging advanced materials and novel technologies. Attention is paid to those newly discovered or applied materials with potential to solve pressing societal problems and improve quality of life, corresponding to environmental protection, medicine, communications, energy, transportation, advanced manufacturing, and related areas.

The series takes into account that, under present strong demands for energy, material, and cost savings, as well as heavy contamination problems and worldwide pandemic conditions, the area of emerging materials and related scalable technologies is a highly interdisciplinary field, with the need for researchers, professionals, and academics across the spectrum of engineering and technological disciplines. The main objective of this book series is to attract more attention to these materials and technologies and invite conversation among the international R&D community.

Assessment of Polymeric Materials for Biomedical Applications
Edited by Vijay Chaudhary, Sumit Gupta, Pallav Gupta, and Partha Pratim Das

Nanomaterials for Sustainable Energy Applications
Edited by Piyush Kumar Sonkar and Vellaichamy Ganesan

Materials Science to Combat COVID-19
Edited by Neeraj Dwivedi and Avanish Kumar Srivastava

Two-Dimensional Nanomaterials for Fire-Safe Polymers
Yuan Hu and Xin Wang

3D Printing and Bioprinting for Pharmaceutical and Medical Applications
Edited by Jose Luis Pedraz Muñoz, Laura Saenz del Burgo Martínez, Gustavo Puras Ochoa, and Jon Zarate Sesma

Polymer Processing: Design, Printing and Applications of Multi-Dimensional Techniques
Abhijit Bandyopadhyay and Rahul Chatterjee

Nanomaterials for Energy Applications
Edited by L. Syam Sundar, Shaik Feroz, and Faramarz Djavanroodi

Wastewater Treatment with the Fenton Process: Principles and Applications
Dominika Bury, Michał Jakubczak, Jan Bogacki, Piotr Marcinowski, and Agnieszka Jastrzębska

Mechanical Behavior of Advanced Materials: Modeling and Simulation
Edited by Jia Li and Qihong Fang

For more information about this series, please visit: www.routledge.com/Emerging-Materials-and-Technologies/book-series/CRCEMT

Wastewater Treatment with the Fenton Process

Principles and Applications

Dominika Bury
Michał Jakubczak
Jan Bogacki
Piotr Marcinowski
Agnieszka Jastrzębska

CRC
CRC Press
Taylor & Francis Group
Boca Raton London New York
CRC Press is an imprint of the
Taylor & Francis Group, an informa business

Designed cover image: Shutterstock

First edition published 2024
by CRC Press
2385 NW Executive Center Drive, Suite 320, Boca Raton FL 33431

and by CRC Press
4 Park Square, Milton Park, Abingdon, Oxon, OX14 4RN

CRC Press is an imprint of Taylor & Francis Group, LLC

ISBN: 978-1-032-35901-4 (hbk)
ISBN: 978-1-032-42742-3 (pbk)
ISBN: 978-1-003-36408-5 (ebk)

DOI: 10.1201/9781003364085

Typeset in Times
by MPS Limited, Dehradun

The Open Access version of Chapters 3, 4, and 9 are funded by Inicjatywa Doskonalosci - Uczelnia badawcza.

Contents

Contents

1 Introduction

The presence of refractory organic compounds in wastewater is a many-year global problem. Despite the extensive development of various wastewater treatment technologies, they still fail to decompose potentially harmful and dangerous compounds. There is an essential need to develop novel solutions that are efficient and cheap. One of the effective methods to reduce pollutants is advanced oxidation processes (AOPs), consisting of effective radical generation, and deserves special attention. This study presents an insight into the used heterogeneous Fenton processes, which is one of the AOPs methods. The application of iron-based heterogeneous catalysts for wastewater treatment and pollution decomposition is promising, as demonstrated by many laboratory, field-based, pilot and full-scale studies. Further, fundamentals, primary applications, crucial process parameters, kinetics, and selected economic aspects were discussed.

The literature review has shown that the iron-based heterogeneous Fenton process can be suitable for wastewater treatment, including industrial wastewater characterized by complex composition. In addition, high efficiency and environmental friendliness indicate the potential for industrial application.

Clean and sufficient quality water is a crucial component of the environment and a resource in industry, agriculture, and human life. WHO and UNICEF noticed that in 2017 more than 785 million people did not have access to at least essential water services in the last their research published. Furthermore, more than 884 million people did not have safe water to drink. These problems affect especially poverty country with not enough developed systems of clean water or its complete absence. Access to safe drinking water is a basic human right, yet many people around the world do not have access to it. This lack of access is often due to poverty, inadequate infrastructure, and environmental degradation. Without safe drinking water, people may be forced to drink from contaminated sources, which can lead to waterborne illnesses such as diarrhea, cholera, and typhoid fever. These illnesses can be particularly dangerous for young children, pregnant women, and those with weakened immune systems. In addition to the health impacts, the lack of safe drinking water can also have economic and social consequences. Without access to safe water, people may have to spend a significant amount of time and energy collecting water, which can limit their ability to attend school or earn a living. Furthermore, the lack of safe water can contribute to conflict and displacement, as people may be forced to move in search of water. In developed countries, problems with clean water mainly generated pollutants from the transfer and agriculture or overexploitation of them. Water pollution is mostly caused by the discharge of untreated essential domestic and industrial wastewater. More than 400 million tons of pollutants are released into water supplies. Moreover, problems with water quality could be generated during accidents and failures in large companies [1–5].

DOI: 10.1201/9781003364085-1

China, India, and some states in the USA and European countries are the places that generated the largest amounts of sewage (Figure 1.1). However, even treated wastewater is not free from pollutants, which are often present at the trace level. This is the effect of the high concentration of residents and the developed industry, including textiles, chemicals, and pharmaceuticals. The wastewater is often a by-product, generated, among others, when washing production lines. The presence of industrial contaminants in ground waters, surface waters, seawater, wastewater treatment plants, soils, and sludges was noticed by scientists from all world [6].

Therefore, it is essential to implement effective wastewater treatment processes that can remove a wide range of pollutants, including those that are present at trace levels. Furthermore, it is crucial to implement regulations and policies that promote the sustainable management of wastewater and the reduction of pollution at the source. This will require the cooperation of governments, industries, and individuals to ensure that the world's water resources are protected for future generations [7]. Industrial wastewater can be dangerous due to the presence of various pollutants that can be harmful to human health and the environment. These pollutants can include chemicals, heavy metals, pathogens, and organic compounds that can have adverse effects on aquatic life, soil, and vegetation. For example, some industrial wastewater contains toxic heavy metals such as lead, mercury, and cadmium, which can accumulate in the food chain and pose a serious threat to human health. Exposure to these heavy metals can result in neurological disorders, kidney damage, and cancer. Furthermore, industrial wastewater can contain organic compounds such as benzene, toluene, and trichloroethylene, which are harmful to human health even at low levels of exposure. These compounds can cause skin and eye irritation, and respiratory problems, and long-term exposure can lead to cancer.

In addition, industrial wastewater can contain pathogenic microorganisms such as bacteria, viruses, and parasites, which can cause waterborne diseases such as cholera, typhoid, and hepatitis A.

Moreover, the discharge of untreated or poorly treated industrial wastewater can have serious environmental consequences. For example, the presence of nutrients in wastewater can cause eutrophication, which can result in the depletion of oxygen in water bodies and the death of aquatic life.

The industrial compounds contaminate the high source of pharmaceuticals, hormones, consumer product chemicals, and other organic wastewater compounds. All of these solutions could be dangerous for the environment and people's health. Much of the compounds accumulate very well in the end, consequently, an increase in the concentration of pollutants in individual components of the environment. The highest threat is related to ones that possess toxic, mutagenic, and carcinogenic properties. They are usually difficult to remove and act as endocrine disruptors [8–11]. Interestingly, hormones presented in water could be linked to an increase in elevated female:male ratios, and their physiological alterations. Moreover, antibiotics could change the microbial structure and nitrogen biotransformation processes [7]. Pollution of wastewater origin can easily get into groundwater and contaminate drinking water resources [12,13]. What is especially important, such

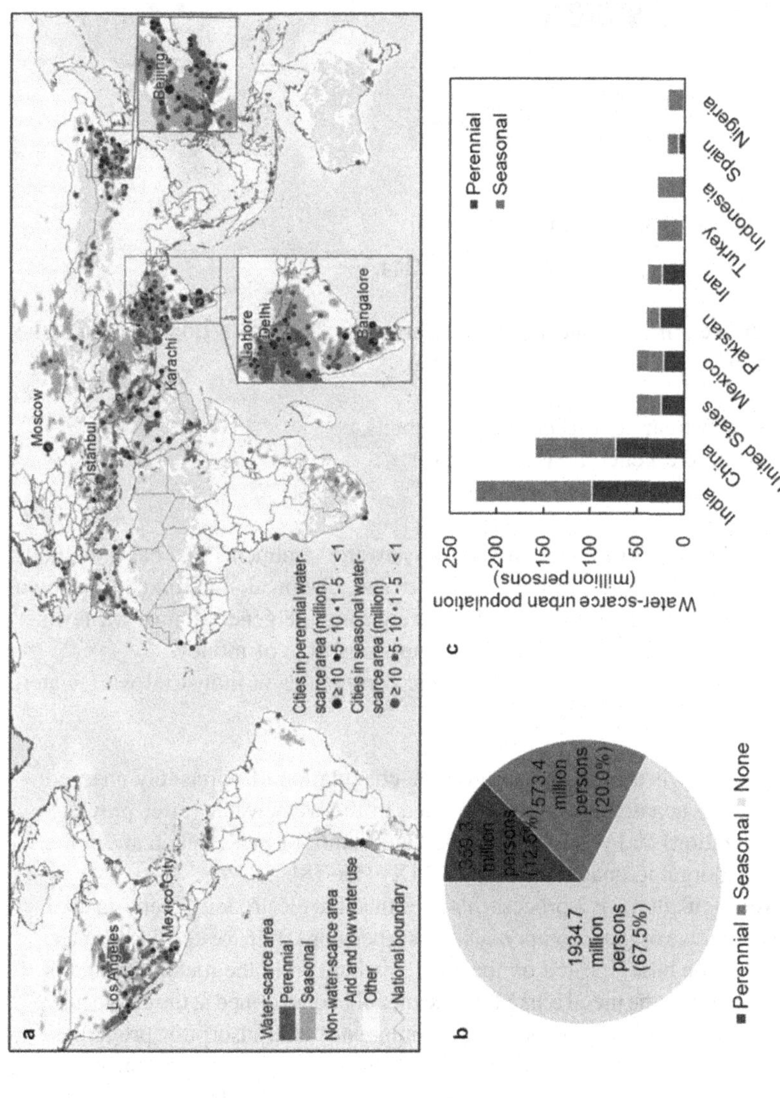

FIGURE 1.1 Current urban water scarcity.

Source: https://www.nature.com/articles/s41467-021-25026-3/figures/1.

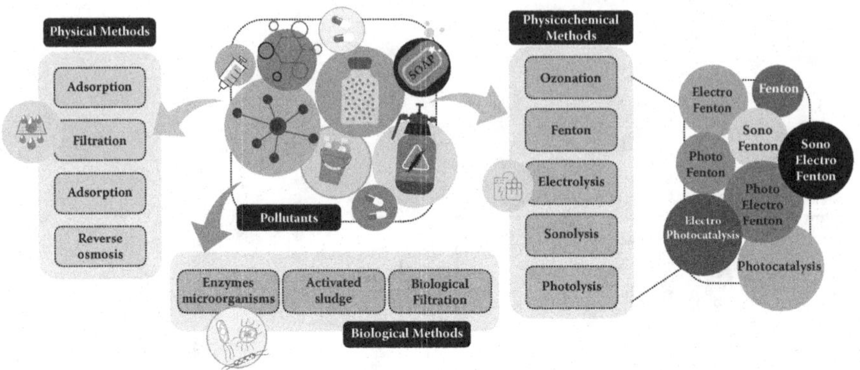

FIGURE 1.2 Kinds of wastewater treatment.

Source: https://www.mdpi.com/2073-4344/12/3/344.

effects appear even in low concentrations. Currently, the basic actions to protect the environment are:

- •. Wastewater treatment to remove pollutants,
- •. Decreasing the generation of wastewater,
- •. Reusing/closing circuits [14].

The most popular and common method is wastewater treatment. The basic methods of wastewater treatment are mechanical, biological, chemical, and mixed, as shown in Figure 1.2 [15]. Other methods, such as decreasing the generation of wastewater are less effective, while reusing/closing circuits cost lots of money.

Due to the usually high content of organic compounds in industrial wastewater, methods such as:

- Coagulation – is a process based on the coagulation of oppositely charged compounds, negative colloids contained in the wastewater (after previous destabilization) and positively charged aluminum or Fe^{3+}, which are in the form of inorganic coagulants (PAC, PIX) [16–18].
- Electrocoagulation – is a process that uses metallic electrodes. Scientists used them to remove organic compounds and suspensions from treated wastewater. The electrode base, located on the anode, is built from the metal part and is used to generate the metal ions M^{n+}. The metallic ions formed at the anode may undergo hydrolysis or form polymeric compounds with adsorption properties.
- Flotation – is the process of separating the fats and flocs present in the wastewater in a flotation device (open separation tank). The process takes place under the influence of dissolved air. Compounds contained in sewage are mixed with fine air bubbles with water and air (a mixture is formed). While mixing, air bubbles stick to the fluff and raise it to the surface. Too heavy substances in the wastewater do not flow out with the bubbles; they settle and then are discharged through the drainage system [17,19].

- Activated carbon adsorption – is a process based on the electron donor-acceptor complex mechanism. These processes reals on pollutants of the adsorbate act as the electron acceptors and the basic sites on the carbon surface serve as the donors. Moreover, scientists noticed the – dispersion interactions and the solvent effects during the adsorption process [20,21].

However, these methods used alone are not effective enough. Treatment installation is required, consisting of more than one unit process. On the other hand, each method has its drawbacks. For example, the filtration requires maintenance of the filter in a proper condition due to pores clogging, which affects less flow velocity, and sorption surface [22]. A by-product of sedimentation is sludge, and the process itself is used to complement a proper treatment process. Additionally, sludge needs to be removed after the process. During the coagulation, it is also necessary to introduce additional chemical compounds, coagulants, and flocculants, which increase the salinity of the wastewater [18,23]. The poor removal of fine emulsions characterizes the adsorption process [24]. In the case of biological wastewater treatment, one should consider the high sensitivity to the content and pH of the toxic compound of the wastewater and the conditions that the process should meet [25]. Most of the methods mentioned above have their weaknesses and drawbacks, which may significantly hinder their usage in industrial wastewater treatment. A growing amount of wastewater encourages to development of new treatment methods and processes [26].

A direction that seems particularly promising is the development of AOPs. The reaction is driven due to the generation of molecules with unpaired valence electrons, so-called highly reactive radicals. Radical addition is the attachment of a radical to a neutral molecule, during which the multiple bonds are broken, resulting in a new product with one unpaired electron [27]. The reaction consists of taking hydrogen atoms from C–H, N–H, or O–H. In the next stage, OH bonds transform into unsaturated C=C bonds or aromatic rings [28]. In AOP, the typically generated radical is the hydroxyl one – $^{\bullet}OH$, one of the strongest known oxidants [29].

The advantages of the process include:

- Lack of specialized conditions and devices necessary to carry out the reaction – "room conditions",
- The reagents for carrying out the process are widely available, thus making the process inexpensive,
- Efficiency and effectiveness of converting organic pollutants into simple non-toxic compounds,
- In the case of heterogeneous catalysts (iron minerals), there is a possibility of reusing the material – saving raw material and activities conducive to environmental protection and reduction of process costs [30].

AOPs differ from each other in radicals generated, the reagents used or required conditions, or other process parameters. The advantages of AOPs include the use of catalysts and oxidants, which are environmentally friendly, and the high efficiency in the oxidation of organic compounds. AOPs are an excellent alternative to hazardous oxidation processes and toxic inorganic oxidants such as permanganates or chlorine. Due to their high efficiency and effectiveness, the AOPs are readily used in the treatment of water and wastewater (including from various industries, including textiles or food). Moreover, it is popular to remove pollutants, during soil and sediment reclamation, municipal sediment conditioning, and volatile organic compounds removal [29].

AOPs processes are divided into:

- Ozone-based AOPs: This type of AOP involves the use of ozone gas (O_3) to generate hydroxyl radicals. Ozone is a strong oxidizing agent that can break down many organic and inorganic pollutants in water. Examples of ozone-based AOPs include ozone alone (O_3), ozone combined with hydrogen peroxide (O_3/H_2O_2), and ozone combined with ultraviolet light (O_3/UV).
- Hydrogen peroxide-based AOPs: This type of AOP involves the use of hydrogen peroxide (H_2O_2) to generate hydroxyl radicals. Hydrogen peroxide is a relatively weak oxidizing agent that requires a catalyst or another oxidizing agent (such as ozone or UV light) to generate hydroxyl radicals. Examples of hydrogen peroxide-based AOPs include Fenton's reagent (Fe^{2+}/H_2O_2) and photo-Fenton's reagent (Fe^{2+}/H_2O_2/UV).
- UV-based AOPs: This type of AOP involves the use of ultraviolet (UV) light to generate hydroxyl radicals. UV light can break down some organic pollutants, but it is not strong enough to break down many inorganic pollutants. Therefore, UV-based AOPs are often combined with other oxidizing agents such as hydrogen peroxide or ozone. Examples of UV-based AOPs include UV alone (UV), UV combined with hydrogen peroxide (UV/H_2O_2), and UV combined with ozone (UV/O_3).
- Electrochemical AOPs: This type of AOP involves the use of an electric current to generate hydroxyl radicals. Electrochemical AOPs can be used to treat a wide range of pollutants, including those that are difficult to treat using other AOPs. Examples of electrochemical AOPs include electro-Fenton (Fe^{2+}/H_2O_2/DC), anodic oxidation (AO), and cathodic reduction (CR) [31].

Another breakdown of AOPs includes:

- Chemical oxidation *vs.* photochemical oxidation: AOPs can be divided into chemical oxidation and photochemical oxidation processes. Chemical oxidation processes use chemical oxidants such as ozone, hydrogen peroxide, or persulfate to generate hydroxyl radicals. Photochemical oxidation processes use UV light irradiation or other light sources to generate hydroxyl radicals.
- Homogeneous *vs.* heterogeneous AOPs: Homogeneous AOPs have the catalyst in the same phase as the solution, whereas heterogeneous AOPs have the catalyst in a different phase. Examples of homogeneous AOPs include Fenton's reagent, photo-Fenton, and the TiO_2-based photocatalytic

process. Examples of heterogeneous AOPs include photocatalysis using immobilized catalysts, such as TiO_2 on glass plates or ceramic beads.
- Traditional *vs.* novel AOPs: Traditional AOPs have been in use for several decades, while novel AOPs are relatively new technologies that have been developed in recent years. Examples of traditional AOPs include Fenton's reagent, ozonation, and photocatalysis. Examples of novel AOPs include plasma-activated water, ultrasound, and sonication.
- Ozone-based *vs.* hydrogen peroxide-based AOPs: Ozone and hydrogen peroxide are two common oxidants used in AOPs. Ozone-based AOPs generate hydroxyl radicals through the reaction of ozone with water, whereas hydrogen peroxide-based AOPs generate hydroxyl radicals through the reaction of hydrogen peroxide with iron (Fe^{2+}) or light.
- Thermal *vs.* non-thermal AOPs: Thermal AOPs rely on heat to generate hydroxyl radicals, whereas non-thermal AOPs do not. Examples of thermal AOPs include wet air oxidation (WAO) and supercritical water oxidation (SCWO). Non-thermal AOPs include photocatalysis, Fenton's reagent, and ozonation [31,32].

Scientists used several methods to generate radicals, as shown in Figure 1.3, especially homogeneous, heterogeneous Fenton process, photo-Fenton or photocatalysis, and reaction with ozone [30].

The most popular of these methods is the Fenton process, which utilized chemical catalysts (including Fe^{2+}, TiO_2 ions). In the classical Fenton process, the hydroxyl radical ($^{\bullet}OH$) can be generated from the reaction between aqueous ferrous ions and hydrogen peroxide (H_2O_2). In the next step, hydrogen peroxide destroys refractory and toxic organic pollutants in wastewater. The conventional Fenton process realizes

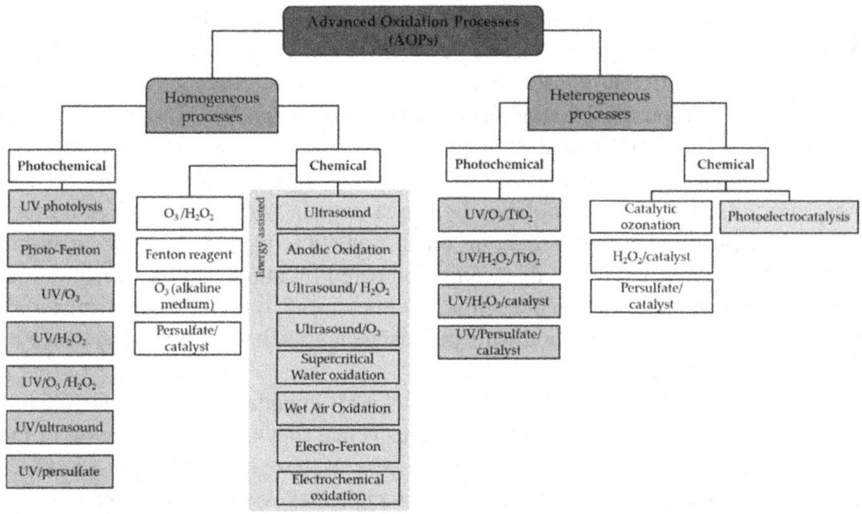

FIGURE 1.3 Types of advanced oxidation processes.

Source: https://www.mdpi.com/2073–4441/11/2/205#.

this by including acid regulation, catalyst mixing, oxidation reaction, neutralization, and solid-liquid separation.

In a homogeneous AOPs, the catalyst and the wastewater or solution to be treated are in the same phase, usually in the liquid phase. In this type of AOPs, the catalyst is typically a metal ion or a combination of metal ions that are used to generate highly reactive hydroxyl radicals that can degrade or mineralize various types of organic and inorganic pollutants in wastewater. One of the most widely used homogeneous AOPs is the Fenton reaction, which involves the reaction of hydrogen peroxide with ferrous ions (Fe^{2+}) to produce hydroxyl radicals. The Fenton reaction is typically carried out at acidic pH conditions, which favors the formation of Fe^{2+} ions. The hydroxyl radicals generated in the Fenton reaction can oxidize a wide range of organic compounds, including aromatic compounds, dyes, and pharmaceuticals. Another example of a homogeneous AOPs is the photo-Fenton reaction, which combines the Fenton reaction with UV light to generate even more hydroxyl radicals. In this process, UV light is used to excite the Fe^{2+} ions, which react with hydrogen peroxide to generate hydroxyl radicals. Homogeneous AOPs have several advantages, including high reaction rates, high efficiency, and the ability to treat a wide range of pollutants. However, they also have some limitations, such as the high cost of the metal catalysts, the requirement for low pH conditions, and the potential for the formation of sludge or by-products that may require further treatment. Additionally, homogeneous AOPs are not effective for treating pollutants that are present in high concentrations or in complex matrices. The homogenous Fenton is defined as the Fenton reaction in which iron salts are used as a catalyst. However, these processes are characterized by many disadvantages. Among them, we distinguish the generation of ferric hydroxide sludge at pH values above 4.0. The sludge has to be removed, which generated additional costs in the process. Moreover, the catalyst is difficult to recycle and reuse. Other problems with the method are high energy consumption and limitation of the operating pH range [33–37].

In a heterogeneous Fenton process, the catalyst and the wastewater or solution to be treated are in different phases. Typically, the catalyst is immobilized on a solid support, such as a porous ceramic or metal oxide material. The wastewater is then passed over or through the catalyst, allowing the catalyst to generate highly reactive hydroxyl radicals that can degrade or mineralize various types of organic and inorganic pollutants. The immobilized catalyst used in heterogeneous Fenton processes is typically based on iron or other transition metals, such as titanium, manganese, or copper. The catalyst can be prepared by impregnation, precipitation, or other methods that involve depositing the metal ions onto the support material. The catalyst can also be modified with various agents, such as carbon, to enhance its reactivity and stability.

Heterogeneous Fenton processes have several advantages over homogeneous Fenton processes. One of the main advantages is that they allow for the reuse of the catalyst, reducing the overall cost of the process. Heterogeneous Fenton processes also have a wider pH range of operation, which makes them more versatile for treating different types of wastewater. Additionally, they do not require the addition of acid to lower the pH, which can result in reduced sludge formation and by-product formation [38,39].

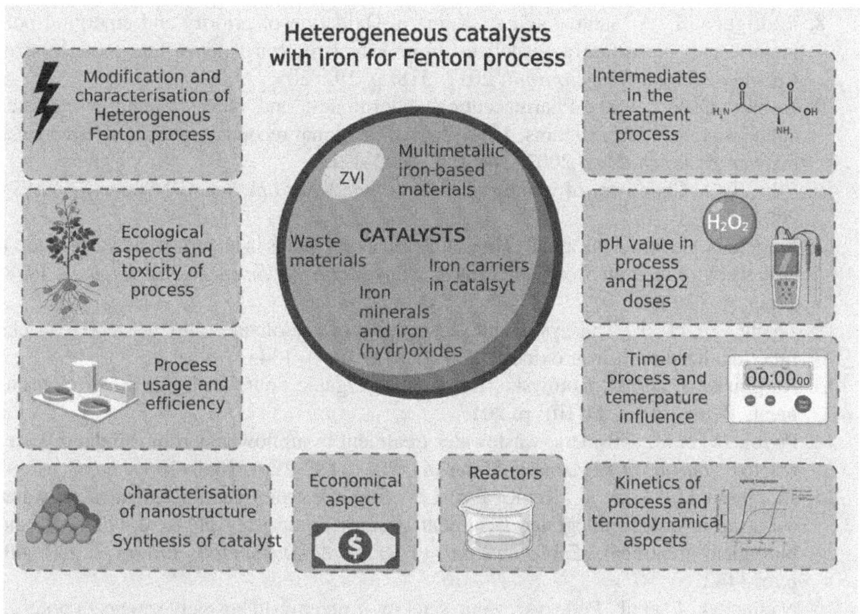

FIGURE 1.4 Summary of the content of the book.

Source: Own work.

Therefore, scientists more often use the heterogeneous Fenton process to degrade wastewater. This study aims to provide brief and summarized information about heterogeneous Fenton processes usage. The secondary aim of the article is to summarize the current state of knowledge regarding the conditions and parameters of the process (Figure 1.4).

REFERENCES

1. Zhang, Y., et al., Nanomaterials-enabled water and wastewater treatment. *NanoImpact*, 2016. **3**.
2. Pulicharla, R., et al., Cosmetic nanomaterials in wastewater: Titanium dioxide and fullerenes. *Journal of Hazardous, Toxic, and Radioactive Waste*, 2016. **20**(1): p. B4014005.
3. Maojun, W., et al., The research on the relationship between industrial development and environmental pollutant emission. *Energy Procedia*, 2011. **5**: p. 555–561.
4. Bury, D., et al., Cleaning the environment with MXenes. *MRS Bulletin*, 2023. **48**: p. 271–282.
5. Jakubczak, M., et al., Excellent antimicrobial and photocatalytic performance of C/GO/TiO_2/Ag and C/TiO_2/Ag hybrid nanocomposite beds against waterborne microorganisms. *Materials Chemistry and Physics*, 2023. **297**: p. 127333.
6. Patel, M., et al., Pharmaceuticals of emerging concern in aquatic systems: Chemistry, occurrence, effects, and removal methods. *Chemical Reviews*, 2019. **119**(6): p. 3510–3673.
7. Schaider, L.A., K.M. Rodgers, and R.A. Rudel, Review of organic wastewater compound concentrations and removal in onsite wastewater treatment systems. *Environmental Science & Technology*, 2017. **51**(13): p. 7304–7317.

8. Rodriguez, S., A. Santos, and A. Romero, Oxidation of priority and emerging pollutants with persulfate activated by iron: Effect of iron valence and particle size. *Chemical Engineering Journal*, 2017. **318**: p. 197–205.
9. Kolpin, D.W., et al., Pharmaceuticals, hormones, and other organic wastewater contaminants in U.S. streams, 1999–2000: A national reconnaissance. *Environmental Science & Technology*, 2002. **36**(6): p. 1202–1211.
10. Ellis, T.G., Chemistry of wastewater. *Encyclopedia of Life Support System (EOLSS)*, 2004. **2**: p. 1–10.
11. Jungclaus, G., V. Avila, and R. Hites, Organic compounds in an industrial wastewater: a case study of their environmental impact. *Environmental Science & Technology*, 1978. **12**(1): p. 88–96.
12. Zhang, Z., et al., The systematic adsorption of diclofenac onto waste red bricks functionalized with iron oxides. *Water*, 2018. **10**: p. 1343.
13. Karpińska, J. and U. Kotowska, Removal of organic pollution in the water environment. *Water*, 2019. **11**(10): p. 2017.
14. Puyol, D., et al., Cosmetic wastewater treatment by upflow anaerobic sludge blanket reactor. *Journal of Hazardous Materials*, 2011. **185**(2): p. 1059–1065.
15. Jhunjhunwala, A., et al., Removal of Levosulpiride from pharmaceutical wastewater using an advanced integrated treatment strategy comprising physical, chemical and biological treatment. *Environmental Progress & Sustainable Energy*, 2020. **40**: p. e13482.
16. Naumczyk, J., et al., Podczyszczanie ścieków z przemysłu kosmetycznego za pomocą procesu koagulacji. *Annual Set The Environment Protection*, 2013: p. 875–891.
17. Bogacki, J.P., et al., Cosmetic wastewater treatment using dissolved air flotation. *Archives of Environmental Protection*, 2017. **43**(No 2).
18. Michel, M., et al., Technological conditions for the coagulation of wastewater from cosmetic industry. *Journal of Ecological Engineering*, 2019. **20**: p. 78–85.
19. Wilinski, P., et al., Pretreatment of cosmetic wastewater by dissolved ozone flotation (DOF). *Desalination and Water Treatment*, 2017. **71**: p. 95–106.
20. Chaiwichian, S. and S. Lunphut, Development of activated carbon from parawood using as adsorption sheets of organic dye in the wastewater. Materials Today: Proceedings, 2021.
21. Liu, Q.-S., et al., Adsorption isotherm, kinetic and mechanism studies of some substituted phenols on activated carbon fibers. *Chemical Engineering Journal*, 2010. **157**(2): p. 348–356.
22. Milichovsky, M., et al., Cellulosic sorption filter materials with surface flocculation activity—a hopeful anticipation of water purification. *Journal of Water Resource and Protection*, 2014. **06**: p. 165–176.
23. Dutta, K., et al., Chemical oxidation of methylene blue using a Fenton-like reaction. *J Hazard Mater*, 2001. **84**(1): p. 57–71.
24. Georgi, A. and F.-D. Kopinke, Interaction of adsorption and catalytic reactions in water decontamination processes: Part I. Oxidation of organic contaminants with hydrogen peroxide catalyzed by activated carbon. *Applied Catalysis B: Environmental*, 2005. **58**: p. 9–18.
25. Sarayu, K. and S. Sandhya, Current technologies for biological treatment of textile wastewater–a review. *Applied Biochemistry and Biotechnology*, 2012. **167**(3): p. 645–661.
26. Wang, D., M. Ji, and C. Wang, Degradation of organic pollutants and characteristics of activated sludge in an anaerobic/anoxic/oxic reactor treating chemical industrial wastewater. *Brazilian Journal of Chemical Engineering*, 2014. **31**: p. 703–713.
27. McMurry, J., Organic chemistry. 7th ed. International student ed. ed. 2008, Belmont, CA: Thomson Brooks/Cole.

28. Jack, R.S., et al., A review of iron species for visible-light photocatalytic water purification. *Environ Sci Pollut Res Int*, 2015. **22**(10): p. 7439–7449.
29. Macías-Quiroga, I.F., et al., Bibliometric analysis of advanced oxidation processes (AOPs) in wastewater treatment: Global and Ibero-American research trends. *Environmental Science and Pollution Research*, 2021. **28**(19): p. 23791–23811.
30. Pereira, M.C., L.C.A. Oliveira, and E. Murad, Iron oxide catalysts: Fenton and Fentonlike reactions – a review. *Clay Minerals*, 2012. **47**(3): p. 285–302.
31. Ismail, G.A. and H. Sakai, Review on effect of different type of dyes on advanced oxidation processes (AOPs) for textile color removal. *Chemosphere*, 2022. **291**: p. 132906.
32. Saien, J., et al., Homogeneous and heterogeneous AOPs for rapid degradation of Triton X-100 in aqueous media via UV light, nano titania hydrogen peroxide and potassium persulfate. *Chemical Engineering Journal*, 2011. **167**(1): p. 172–182.
33. Muruganandham, M., et al., Recent developments in homogeneous advanced oxidation processes for water and wastewater treatment. *International Journal of Photoenergy*, 2014. **2014**: p. 821674.
34. Rizzo, L., Addressing main challenges in the tertiary treatment of urban wastewater: Are homogeneous photodriven AOPs the answer? *Environmental Science: Water Research & Technology*, 2022. **8**(10): p. 2145–2169.
35. Anandan, S., V. Kumar Ponnusamy, and M. Ashokkumar, A review on hybrid techniques for the degradation of organic pollutants in aqueous environment. *Ultrasonics Sonochemistry*, 2020. **67**: p. 105130.
36. Peng, W., et al., Non-radical reactions in persulfate-based homogeneous degradation processes: A review. *Chemical Engineering Journal*, 2021. **421**: p. 127818.
37. Jakubczak, M., et al. Multifunctional carbon-supported bioactive hybrid nanocomposite (C/GO/NCP) bed for superior water decontamination from waterborne microorganisms. *RSC Advances*, 2021. **11**: p. 18509–18518.
38. Luo, H., et al., Application of iron-based materials in heterogeneous advanced oxidation processes for wastewater treatment: A review. *Chemical Engineering Journal*, 2021. **407**: p. 127191.
39. Wang, D., et al., Perspectives on surface chemistry of nanostructured catalysts for heterogeneous advanced oxidation processes. *Environmental Functional Materials*, 2022. **1**(2): p. 182–186.

2 Fenton Process Principle

PROCESS PRINCIPLE

The Fenton process is one of the advanced oxigen processes (AOPs), discovered in 1894 [1,2]. In the Fenton process, as in any AOP, radicals are formed. Radicals (formerly free radicals) are atoms or molecules containing unpaired electrons. The Fenton process consists of two independent stages: the Fenton reaction and coagulation. The first step of the Fenton process, the Fenton reaction, involves oxidizing Fe^{2+} to Fe^{3+} while reducing hydrogen H_2O_2 under acidic conditions [3–5] to form hydroxyl radicals that could break down almost all known organic pollutants [6]. In the second stage, coagulation, the pH is raised, usually to alkaline, resulting in the breakdown of unreacted H_2O_2, oxidation of Fe^{2+} to Fe^{3+}, and coagulation of iron. A diagram of the Fenton process is shown in Figure 2.1.

FENTON REACTION

The reaction (Eq. 2.1) is known as the Fenton reaction. The Fenton reaction is the catalytic breakdown of hydrogen peroxide into a hydroxyl ion and a hydroxyl radical, the catalyst being a divalent iron ion. Subsequently, a series of further reactions occur. The most important reactions cited in many scientific publications include, *i.a.*, (Eqs. 2.2–2.5).

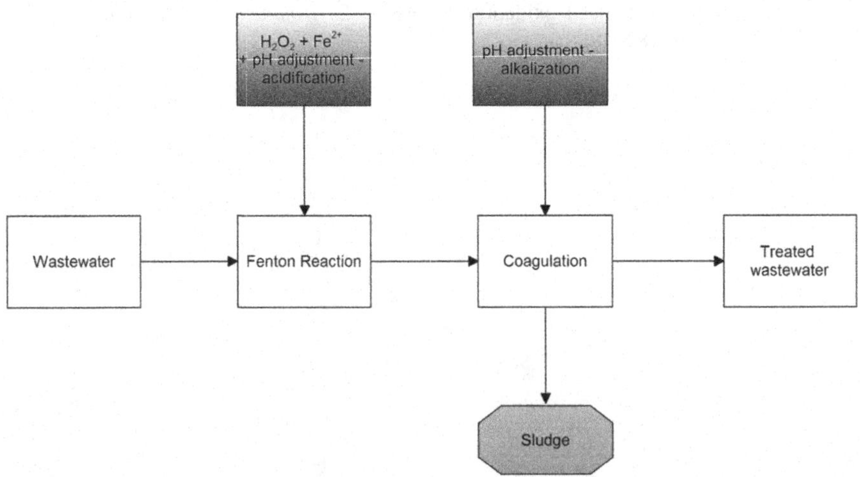

FIGURE 2.1 Diagram of the Fenton process.

Source: Own elaboration.

DOI: 10.1201/9781003364085-2

$$Fe^{2+} + H_2O_2 \rightarrow Fe^{3+} + OH^- + HO^\bullet \quad (2.1)$$

$$Fe^{3+} + H_2O_2 \rightarrow Fe^{2+} + HO_2^\bullet + H^+ \quad (2.2)$$

$$Fe^{2+} + HO^\bullet \rightarrow Fe^{3+} + OH^- \quad (2.3)$$

$$Fe^{2+} + HO_2^\bullet \rightarrow Fe^{3+} + HO_2^- \quad (2.4)$$

$$Fe^{3+} + HO_2^\bullet \rightarrow Fe^{2+} + O_2 + H^+ \quad (2.5)$$

The Fenton reaction is a radical reaction in which radicals are formed. Radical reactions are chain reactions and proceed in three stages:

- Initiation-formation of radicals,
- Propagation-reaction in which the number of radicals does not change,
- Termination-recombination of two radicals into a non-radical product or radical quenching combined with a redox reaction.

Reactions 2.1 and 2.2 should be classified as initiation and reactions 2.3–2.5 as termination. A thorough understanding of the reaction mechanism requires an independent look at each stage and the presentation of the reactions taking place.

Initiation

In addition to reactions 2.1 and 2.2, which form the core chemistry of the Fenton process, the following reactions should also be indicated [7,8].

Formation of hydroxyl radicals ($HO^\bullet$) (Eqs. 2.6–2.7):

$$Fe^{2+} + H_2O_2 \rightarrow Fe^{3+} + OH^- + HO^\bullet \quad (2.6)$$

$$FeOH^{2+} \rightarrow Fe^{2+} + HO^\bullet \quad (2.7)$$

Formation of a hydroperoxyl radical ($HO_2^\bullet$) (Eqs. 2.8–2.9):

$$Fe^{3+} + H_2O_2 \rightarrow Fe^{2+} + HO_2^\bullet + H^+ \quad (2.8)$$

$$FeOH^{2+} + H_2O_2 \rightarrow Fe^{2+} + HO_2^\bullet + H_2O \quad (2.9)$$

Formation of a superoxide anion radical (${}^\bullet O_2^-$) (Eqs. 2.10–2.11):

$$Fe^{3+} + H_2O_2 \rightarrow Fe^{3+} + {}^\bullet O_2^- + 2H^+ \quad (2.10)$$

$$FeOH^{2+} \rightarrow Fe^{2+} + HO^{\bullet} \quad (2.11)$$

Propagation

As a result of the propagation reaction, radicals are transformed, and radicals with less and less redox potential will be transformed into a form [7,9]. Hydroperoxy radicals ($^{\bullet}OOH$) exhibit a lower oxidizing potential than $^{\bullet}OH$ (E_0 = 1.65 V *vs.* 2.8 V) (see Table 2.1) and react with iron or iron ions. Combined or disproportionate molecules form peroxide, anionic radical $^{\bullet}O_2^-$ and hydrogen ion (H^+) (Eqs. 2.12–2.16) [8,10,11]:

$$HO^{\bullet} + H_2O_2 \rightarrow HO_2^{\bullet} + H_2O \quad (2.12)$$

$$HO^{\bullet} + HO_2^- \rightarrow HO_2^{\bullet} + HO^- \quad (2.13)$$

$$HO_2^{\bullet} + H_2O_2 \rightarrow HO^{\bullet} + H_2O + O_2 \quad (2.14)$$

$$HO_2^{\bullet} + H_2O \rightarrow O_2^{\bullet -} + H_3O^+ \quad (2.15)$$

$$O_2^{\bullet -} + H_2O_2 \rightarrow HO^{\bullet} + HO^- + O_2 \quad (2.16)$$

TABLE 2.1
Types of Oxidants (Self-Prepared)

Oxidizing Agent	Potential [V]	Oxidizing Agent	Potential [V]
$^{\bullet}OH$	2.76	H_2O_2 (acidic)	1.78
Oxygen atom (O)	2.42	H_2O_2 (alkaline)	0.85
O_3 (acidic)	2.07	ClO_2	1.71
O_3 (neutral)	1.24	$HO^{\bullet}_2$	1.70
F_2	2.87	Cl_2	1.36
Pb^{4+}	1.67	Au^+	1.69
Ag^+	0.80	$S_2O_8^{2-}$	2.05
PbO_2	1.69	HClO	1.63
Ce^{4+}	1.61	Au^{3+}	1.4
Cl_2	1.48	N_2O	1.76
Am^{4+}	2.60	K_4XeO_6	8.0
$H_2N_2O_2$	2.65	H_4XeO_6	2.42
Cm^{4+}	3.0	XeO_3	2.10
XeF	3.4	Tb^{4+}	3.1
Cf^{4+}	3.3	Pr^{4+}	3.2
Cu^{3+}	2.4	FeO_4^{2-}	2.20
F_2O	2.153	$S_2O_8^{2-}$	2.123
$HFeO_4^-$	2.09	OH	2.02
Cu_2O_3	2.0	Ag^{2+}	1.980
Co^{3+}	1.83	Ag_2O_2	1.802

Termination

Termination reactions stop the chain reaction [7,9,12].

Recombination of two radicals into a non-radical product (Eqs. 2.17–2.19):

$$HO^{\bullet} + HO_2^{\bullet} \rightarrow H_2O + O_2 \quad (2.17)$$

$$HO^{\bullet} + HO^{\bullet} \rightarrow H_2O_2 \quad (2.18)$$

$$HO_2^{\bullet} + HO_2^{\bullet} \rightarrow H_2O_2 + O_2 \quad (2.19)$$

Radical quenching combined with redox reaction (Eqs. 2.20–2.27):

$$Fe^{2+} + HO^{\bullet} \rightarrow Fe^{3+} + HO^- \quad (2.20)$$

$$Fe^{2+} + HO_2^{\bullet} \rightarrow Fe^{3+} + HO_2^- \quad (2.21)$$

$$Fe^{3+} + {}^{\bullet}O_2^- \rightarrow Fe^{2+} + O_2 \quad (2.22)$$

$$Fe^{3+} + HO_2^{\bullet} \rightarrow Fe^{2+} + O_2 + H^+ \quad (2.23)$$

$$Fe^{2+} + HO_2^{\bullet} + H^+ \rightarrow Fe^{3+} + H_2O_2 \quad (2.24)$$

$$Fe(OH)^{2+} + HO_2^{\bullet} \rightarrow Fe^{2+} + O_2 + H_2O \quad (2.25)$$

$$Fe(OH)^{2+} + {}^{\bullet}O_2^- \rightarrow Fe^{2+} + O_2 + HO^- \quad (2.26)$$

$$H^+ + HO^{\bullet} + e^- \rightarrow H_2O \quad (2.27)$$

In the Fenton reaction in the absence of organic substrates, excess Fe^{2+} ions bind ${}^{\bullet}OH$ [13].

SUBREACTION OF THE FENTON PROCESS

Reactions with Organic Compounds

The Fenton reaction is used in practice, *i.a.* in wastewater treatment. Organic compounds contained in wastewater, mainly dissolved, undergo radical attack and oxidation [2,14,15]. The mechanism of radical attack is, *i.a.*, dehydrogenation and hydroxylation. In the following reactions, R stands for saturated organic compound, $R^{\bullet}$ stands for organic radical, and R^+ stands for organic carbocation [13].

Organic compounds can be hydrolyzed under the influence of water, according to the reaction. However, hydrolysis is usually ineffective, so most compounds remain in their initial form. The hydroxyl radical breaks the hydrogen atom from the organic substrate (RH), forming an organic radical ($R^{\bullet}$) [6]. An organic radical ($R^{\bullet}$) reacts rapidly with a dissolved oxygen molecule to form an organic peroxide radical ($ROO^{\bullet}$) [6]. Organic peroxide radical ($ROO^{\bullet}$) removes the hydrogen atom from the organic substrate, forming organic hydroperoxide (ROOH) and one more organic radical ($R^{\bullet}$) [6]. Hydroxyl radicals can oxidize organic compounds by electron transfer reaction (as an oxidizer), detachment of protons to form organic radicals ($R^{\bullet}$) and electrophilic addition to a double bond or aromatic ring [10,11,16]. Organic radicals ($R^{\bullet}$) react with hydrogen peroxide, iron or iron ions, dissolved oxygen, water, and hydroperoxyl radicals ($HOO^{\bullet}$) or form a dimer [8,17,18]. Detailed responses are described below (Eqs. 2.28–2.49):

$$R + HO^{\bullet} \rightarrow \text{intermediate products} \rightarrow CO_2 + H_2O \quad (2.28)$$

$$R + Fe^{3+} \rightarrow Fe^{2+} + R^{+} \quad (2.29)$$

$$R^{+} + H_2O \rightarrow ROH + H^{+} \quad (2.30)$$

$$RH + HO^{\bullet} \rightarrow R^{\bullet} + H_2O \quad (2.31)$$

$$RH + HO^{\bullet} \rightarrow RH(OH^{\bullet}) \quad (2.32)$$

$$R^{\bullet} + O_2 \rightarrow ROO^{\bullet} \quad (2.33)$$

$$ROO^{\bullet} + RH \rightarrow ROOH + R^{\bullet} \quad (2.34)$$

$$R^{\bullet} + H_2O_2 \rightarrow ROH + OH^{\bullet} \quad (2.35)$$

$$R^{\bullet} + Fe^{3+} \rightarrow R^{+} + Fe^{2+} \quad (2.36)$$

$$R^{\bullet} + Fe^{2+} \rightarrow R^{-} + Fe^{3+} \quad (2.37)$$

$$R^{\bullet} + O_2 \rightarrow RO_2^{\bullet} \quad (2.38)$$

$$RO_2^{\bullet} + H_2O \rightarrow ROH + HO_2^{\bullet} \quad (2.39)$$

$$RO_2^{\bullet} + Fe^{2+} \rightarrow RO_2^{-} + Fe^{3+} \quad (2.40)$$

$$R^{\bullet} + OOH^{\bullet} \rightarrow RO_2H \quad (2.41)$$

$$RO_2H + Fe^{2+} \rightarrow Fe^{3+} + OH^{-} + RO^{\bullet} \quad (2.42)$$

$$RO_2H + Fe^{3+} \rightarrow RO^\bullet + Fe^{2+} + H^+ \quad (2.43)$$

$$2R^\bullet \rightarrow R - R \quad (2.44)$$

$$RO_2^\bullet + RO_2^\bullet \rightarrow ROOOOR \quad (2.45)$$

$$FeO^{2+} + RH \rightarrow Fe^{2+} + ROH \quad (2.46)$$

$$R_1O_2^\bullet + R_2H \rightarrow HO_2^\bullet + R_1R_2 \quad (2.47)$$

$$Fe(RCOO)^{2+} \rightarrow Fe^{2+} + R^\bullet + CO_2 \quad (2.48)$$

$$HO^\bullet + R_1 - C = C - R_2 \rightarrow R_1 - (HO)C - C^\bullet R_2 \quad (2.49)$$

Reactions can also occur in the oxidation process producing iron species [19,20]. During the process, radical reactions also occur (reactions between hydroxyl radicals with inorganic substances or the decomposition of H_2O_2 to water and oxygen) [10,11].

Finally, as a result, compounds unstable in an aqueous solution like tetroxides (ROOOOR) are formed. They decompose into alcohols, ketones, and aldehydes. An example of the attack of hydroxyl radicals on an organic compound, methane, can be presented as follows [58] (Eqs. 2.50–2.54):

$$HO^\bullet + CH_4 \rightarrow {}^\bullet CH_3 + H_2O \quad (2.50)$$

$$HO^\bullet + {}^\bullet CH_3 \rightarrow CH_3OH \quad (2.51)$$

$$CH_3OH + HO^\bullet \rightarrow CH_2O \quad (2.52)$$

$$CH_2O + HO^\bullet \rightarrow HCO_2 + H_2 \quad (2.53)$$

$$HCO_2 + HO^\bullet \rightarrow CO_2 + H_2O \quad (2.54)$$

Reaction in the Presence of Organic Complexing Agents

The production of radicals can be additionally supported by complexing compounds (Eqs. 2.55–2.60):

$$Fe(C_2O_4) + H_2O_2 \rightarrow Fe(C_2O_4)^+ + HO^\bullet + HO^- \quad (2.55)$$

$$[Fe(C_2O_4)_3]^{3-} \rightarrow [Fe(C_2O_4)_2]^{2-} + C_2O_4^{\bullet -} \quad (2.56)$$

$$C_2O_4^{\bullet -} \rightarrow CO_2 + CO_2^{\bullet -} \tag{2.57}$$

$$CO_2^{\bullet -} + O_2 \rightarrow CO_2 + O_2^{\bullet -} \tag{2.58}$$

$$Fe^{3+} + C_2O_4^{\bullet -} \rightarrow Fe^{2+} + 2CO_2 \tag{2.59}$$

$$FeEDTA + H_2O_2 \rightarrow FeOEDTA + H_2O \tag{2.60}$$

Side Reactions

During the first stage of the Fenton process, numerous additional side reactions not related to radical processes also take place [7,9]. They are mostly related to ROS formation and rearrangement (Eqs. 2.61–2.63):

$$H_2O_2 + H_2O_2 \rightarrow 2H_2O + O_2 \tag{2.61}$$

$$Fe^{3+} + O_2^{-} \rightarrow Fe^{2+} + O_2 \tag{2.62}$$

$$H_2O_2 + H_2O \rightarrow HO_2^{-} + H_3O^{+} \tag{2.63}$$

Other Metals

The Fenton process can be carried out not only with the use of Fe^{2+} ions. Any reduced ionic form of the metal can be a potential catalyst. Among the potential ions, one can point out, for example, Cu^{+}, Co^{2+}, and Mn^{2+}, according to exemplary reactions (Eq. 2.64):

$$M^{n+} + H_2O_2 \rightarrow M^{(n+1)+} + HO^{-} + HO^{\bullet} \tag{2.64}$$

where M is metal. Some exemplary reactions could be given here [59] (Eqs. 2.65–2.72):

$$Cu^{+} + H_2O_2 \rightarrow Cu^{2+} + HO^{-} + HO^{\bullet} \tag{2.65}$$

$$Co^{2+} + H_2O_2 \rightarrow Co^{3+} + HO^{-} + HO^{\bullet} \tag{2.66}$$

$$Ce^{3+} + H_2O_2 \rightarrow Ce^{4+} + HO^{-} + HO^{\bullet} \tag{2.67}$$

$$Ce^{4+} + H_2O_2 \rightarrow Ce^{3+} + HO_2^{\bullet} + H^{+} \tag{2.68}$$

$$Mn^{2+} + H_2O_2 \rightarrow Mn^{3+} + HO^{-} + HO^{\bullet} \tag{2.69}$$

$$Mn^{3+} + H_2O_2 \rightarrow Mn^{2+} + HO_2^{\bullet} + H^+ \tag{2.70}$$

$$Mn^{3+} + H_2O_2 \rightarrow Mn^{4+} + HO^- + HO^{\bullet} \tag{2.71}$$

$$Mn^{4+} + H_2O_2 \rightarrow Mn^{3+} + HO_2^{\bullet} + H^+ \tag{2.72}$$

Creation of a catalyst is also possible as a result of inter-metallic action [60] (Eq. 2.73):

$$Fe^{3+} + Cu^+ \rightarrow Cu^{2+} + Fe^{2+} \tag{2.73}$$

Other Oxidants

As with the ion catalyst, the classic oxidizer, hydrogen peroxide, can also be replaced by another. Alternative oxidants include calcium peroxide, magnesium peroxide, percarbonate, persulfate and peroxymonosulfate (PMS), as shown in the following reactions [61,62] (Eqs. 2.74–2.79):

$$S_2O_8^{2-} + e \rightarrow SO_4^{-\bullet} + SO_4^{2-} \tag{2.74}$$

$$S_2O_8^{2-} + HO^{\bullet} \rightarrow SO_4^{-\bullet} + HSO_4^- + O \tag{2.75}$$

$$HSO_5^- \rightarrow SO_4^{-\bullet} + HO^{\bullet} \tag{2.76}$$

$$HSO_5^- + Fe^{2+} \rightarrow Fe^{3+} + SO_4^{-\bullet} + HO^- \tag{2.77}$$

$$CaO_2 + 2H_2O \rightarrow Ca(OH)_2 + H_2O_2 \tag{2.78}$$

$$MgO_2 + 2H_2O \rightarrow Mg(OH)_2 + H_2O_2 \tag{2.79}$$

Additional positive effects may occur when using calcium peroxide. Calcium hydroxide, which is formed by the decomposition of calcium peroxide to hydrogen peroxide, can be used at the final stage of coagulation as a floc mass and an additional alkalizing agent.

CONDITIONS FOR CARRYING OUT THE FENTON REACTION

The Fenton reaction can potentially occur over a wide pH range; however, it is believed that the optimal pH is around 3.0. This is related to the properties of reactants – divalent iron ions and hydrogen peroxide. Hydrogen peroxide is a weak acid. The condition for its activation, breaking the bond between oxygen atoms to produce a hydroxyl ion and a radical, is the reactivity of the molecule. This reactivity decreases significantly when the pH drops below 3.0. When the pH rises to about 5.5 and above, the solubility of the iron begins to decline and the solidification process begins. The catalyst ions are thus removed from the reaction environment. On the other hand, when the pH is further increased to 8.0 and above, in addition to the alkaline coagulation of iron and additionally under alkaline conditions, the decomposition of the weak acid that is hydrogen peroxide occurs.

So under acidic conditions, Fenton's reaction processes are fast and efficient. This is due to the presence of H^+ ions, which are necessary for the decomposition of H_2O_2. Further redox reactions occur in the presence of organic compounds with an excess of Fe^{2+} ions. Of course, the process can be carried out under inert or alkaline conditions, but it is less effective than at low pH due to the formation of stable iron complexes, the formation of iron and iron hydroxides and the decomposition of hydrogen peroxide. The formation of stable complexes hinders oxidation. In addition, increasing the pH reduces the dissociation of complexes [13]. However, there will be no radical mechanism and forms of Fe^{4+} will arise (Eqs. 2.80–2.82) [13]. However, the decomposition of organic compounds is still possible because Fe^{4+} compounds are considered strong oxidants.

$$Fe^{2+} + H_2O_2 \rightarrow (FeO)^{2+} + H_2O \tag{2.80}$$

$$FeO^{2+} + H_2O_2 \rightarrow Fe^{2+} + O_2 + H_2O \tag{2.81}$$

$$FeO^{2+} + Fe^{2+} \rightarrow 2FeOH^{2+} \tag{2.82}$$

The process can continue because the hydroxyl radical can react with compounds adsorbed on flocs (Eq. 2.83) [7]:

$$(RH)ads\backslash RH + {}^{\bullet}OH \rightarrow intermediates \rightarrow CO_2 + H_2O \tag{2.83}$$

FINAL COAGULATION

Divalent iron is usually dosed as an acidic sulfate solution. The disadvantage is that acidification further increases the salinity of the wastewater. After the chemical oxidation step, the chemical reaction is extinguished by alkalization. The purpose of the alkalization step is to stop the radical oxidation reaction. Hydrogen peroxide, as a weak acid, will decompose in an alkaline environment. The higher the pH, the faster the rate of decomposition of hydrogen peroxide will be obtained. In addition, the oxygen formed by the decomposition of hydrogen peroxide will be used to oxidize Fe^{2+} to Fe^{3+}. The solubility of Fe^{2+} forms is higher than Fe^{3+} forms, in this pH range hydroxide forms will prevail, which will initiate the solidification process (Eq. 2.20). The alkalization process is carried out at a pH of about 8.5–9.5, which on the one hand is considered optimal for alkaline coagulation with iron-based coagulants, and on the other hand it is the upper maximum pH allowed for wastewater disposal. Alkalization is a secondary source of the increase in wastewater salinity. Alkalinization has an additional effect (Eqs. 2.82–2.85) [10]:

$$2Fe^{2+} + 2H^{+} + H_2O_2 \rightarrow 2Fe^{3+} + 2H_2O \quad (2.84)$$

$$2Fe^{3+} + 6OH^{-} \rightarrow 2Fe(OH)_3\downarrow \quad (2.85)$$

One of the disadvantages of this process is the formation of iron sludge, which precipitates during pH regulation at the end of the purification process. The slow regeneration of sediment-forming iron ions increases the pH and increases the precipitation of complexes. It promotes the formation of unsolvable complexes and increases the number of hydroxyl compounds. The sludge treatment process is expensive, environmentally hazardous and difficult due to colloidal properties and particle dispersion [21,22]. However, reducing the amount of sludge and maintaining a low pH can accelerate the regeneration of iron ions [23]. During the coagulation stage, two opposing processes occur. On the one hand, a coagulation process takes place, which causes the formation of flocs of iron hydroxide sludge. Thanks to coagulation, including by sorption on the newly formed sediment, it is possible to remove dissolved organic compounds remaining after the Fenton reaction. Sludge herds slowly settle at the bottom of the reactor/settling tank. On the other hand, the decomposition of hydrogen peroxide releases oxygen, which is not fully used for iron oxidation. Excess oxygen, exceeding its small solubility in the liquid, separates from it in the form of floating bubbles. As a result, two opposing streams of sedimenting sediment flocs and floating oxygen bubbles appear. This leads to intensive mixing of wastewater.

REAGENT DOSES

In order to achieve an optimal purification effect, and thus the most effective decomposition of organic compounds, it is necessary to select the right proportions between the reactants used in the peroxide reaction and Fe^{2+} ions and organic compounds contained in wastewater. The configuration of the Fenton process is shown in Figure 2.2.

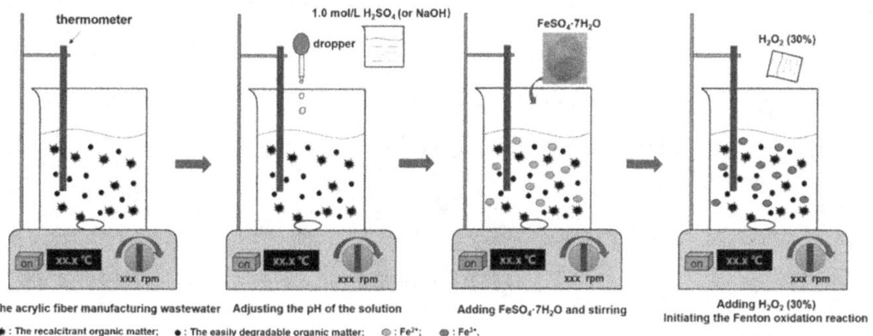

FIGURE 2.2 Fenton process configuration.

Source: https://www.mdpi.com/2073-4441/14/18/2913.

The most commonly used parameter understood as a measure of the content of organic compounds in wastewater is chemical oxygen demand (COD). It is considered that there should be an appropriate mass ratio of COD to oxidizer. Many studies have been conducted and there is no complete agreement among the authors, but the most common view is that the mass ratio H_2O_2/COD should be in the range of 0.5:1 to 4:1. The most common is 1:1. The exact composition of the sample or waste water subjected to the treatment process seems to affect the correct proportion of reagents. There are reports of proportions of up to several hundred to one. In addition, the mass ratio of the oxidizer to the catalyst is important. It is usually considered that it should be H_2O_2/Fe^{2+} in the range from 1:1 to 10:1, most often 4:1. It should be remembered that the amount of Fe^{2+} is responsible not only for the effectiveness of the Fenton reaction, but also the dose of coagulant during the final coagulation. The dose of coagulant needed is very important, due to the possible content of colloids and suspensions in the treated wastewater.

FENTON PROCESS MODIFICATION

Due to the proven and recognized effectiveness of the Fenton process, it is a promising method of wastewater treatment. However, the process is carried out at a low pH, during the process there is a need to adjust the pH twice – initially acidification, then alkalization, which leads to a significant increase in salinity. In addition, the final treatment produces a sludge that requires further processing. The catalyst is removed after the process and there is no simple way to recover it.

For this reason, numerous attempts are made to improve the process, including in particular the elimination of these defects. The modifications to the process used include changing the type of catalyst, its form, method of input, and method of recovery. Research on the transformation of the oxidant is being conducted. In addition, external energy sources are introduced that activate the chemical reactions taking place. Numerous modification options are shown in Figure 2.3.

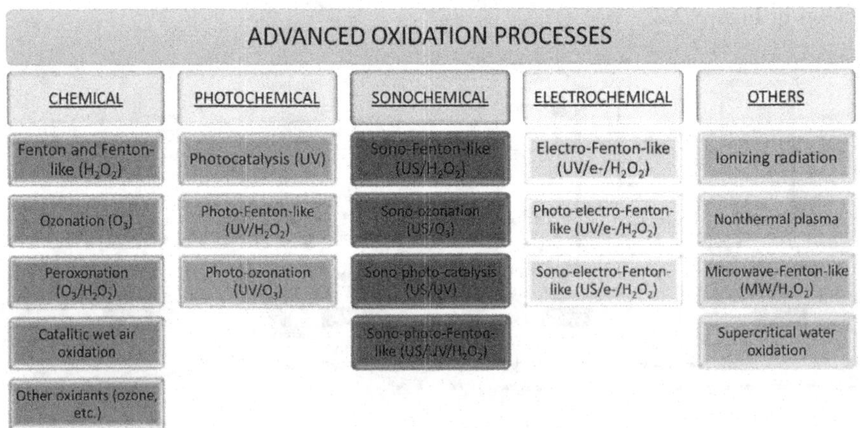

FIGURE 2.3 Modification of the heterogeneous Fenton process.

Source: https://www.mdpi.com/2073-4344/12/4/358.

Pseudo-Fenton Process

The classical Fenton process is a homogeneous process – reactants are dissolved in the aqueous phase, and the reaction environment is single-phase. A possible modification is the introduction of an iron source other than Fe^{2+} ions. A substitute source of iron may be Fe^{3+} ions. To initiate the Fenton reaction, they must first be reduced. A key modification of the Fenton process is the so-called pseudo-Fenton, in which trivalent iron is reduced to divalent iron. After the reduction of Fe^{3+} to Fe^{2+}, this process can be continued by oxidizing Fe^{2+} to Fe^{3+} to form H_2O_2 according to the reactions. It is worth noting that due to the greater reactivity of H_2O_2, the Fenton process is a faster reaction than the pseudo-Fenton reaction (Eqs. 2.41–2.47) [24].

$$Fe^{3+} + H_2O_2 \rightarrow Fe^{2+} + HO_2^{\bullet} + H^+ \tag{2.86}$$

Eq. (2.86) is also known as the pseudo-Fenton reaction. However, this reaction also occurs in the classical Fenton process. The reaction efficiency is high in an acidic environment. As in the classical Fenton reaction, the optimal pH is 3.0 [25].

Photo-Fenton Process

It is possible to modify the Fenton process by using additional energy sources such as electro-Fenton and improved plasma, ultrasound, ultraviolet and photo-Fenton with UV light or sunlight (Figure 2.4).

One of the most popular modifications of the Fenton process involving light is called photo-Fenton (Eq. 2.87) [26]. Photo-Fenton process configuration is shown in Figure 2.4.

$$H_2O_2 + hv \rightarrow {}_2HO^{\bullet} \tag{2.87}$$

An alternative to the UV-light photo-Fenton process can be a solar-Fenton process, which is something more energy-efficient, and cheaper. Light can effectively contribute to the catalytic reproduction of Fe^{2+} and photo-decarboxylation of the ferric carboxylate complex (Eq. 2.88) [27–30].

$$Fe(RCOO)^{2+} + hv \rightarrow Fe^{2+} + CO_2 + R^{\bullet} \tag{2.88}$$

The sono-Fenton process involves ultrasonic treatment with many additional modifications, such as the resulting $^{\bullet}OH$ and H_2O_2 as a cavitation effect, promoting the removal of impurities and reducing the amount of oxidizer added. Additional ultrasound increases the rate of decomposition of H_2O_2 and generation of $^{\bullet}OH$. In addition, it accelerates mass transfer and the reduction rate of Fe^{3+} to Fe^{2+} (Eqs. 2.89–2.91), where))) are ultrasound) [31,32].

$$H_2O +))) \rightarrow {}^{\bullet}OH + {}^{\bullet}H \tag{2.89}$$

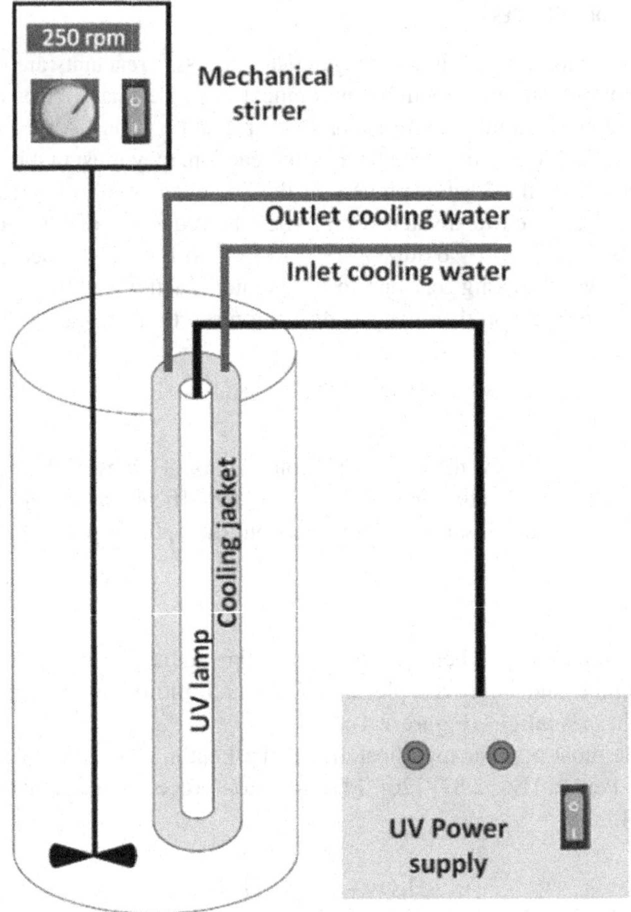

FIGURE 2.4 Photo-Fenton process configuration.

Source: https://www.mdpi.com/2073-4441/11/9/1849.

$$O_2 +))) + 2H_2O \rightarrow 4^{\bullet}OH \tag{2.90}$$

$$H_2O_2 +))) \rightarrow 2^{\bullet}OH \tag{2.91}$$

Another modification of the Fenton process is the use of electric current. The electro-Fenton process configuration is shown in Figure 2.5.

In the electro-Fenton process, Brillas *et al.* [13] used electrodes to produce Fe^{2+} by electrochemical dissolution (type I electro-Fenton process) of electrodes or electrogeneration of hydrogen peroxide (electro-Fenton process type II) [13]. Hydrogen peroxide forms on the cathode in an acidic solution to reduce oxygen by two electrons. At the same time, they are regenerated by anodic reduction of iron ions. When coming into contact, iron and carbon molecules bind to the electrolyte,

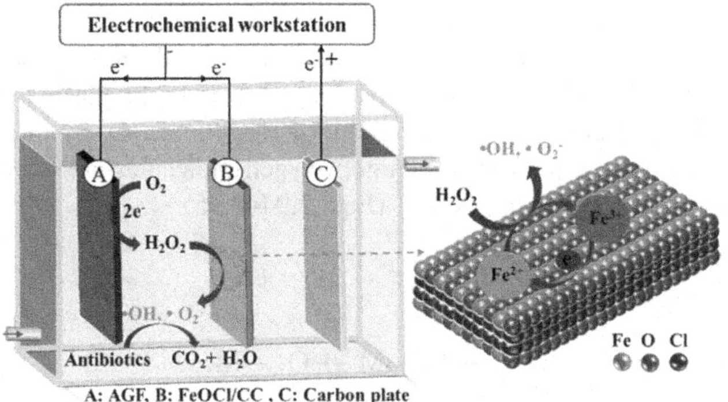

FIGURE 2.5 Electro-Fenton process configuration.

Source: https://pubs.acs.org/doi/10.1021/acssuschemeng.0c08705.

causing the formation of galvanic cells, in which iron acts as an anode and carbon as a cathode. Iron gives two electrons, forming a ferrous ion at the cathode (Eq. 2.92). During the electro-Fenton process, direct oxidation involves transferring the charge of hydroxyl radicals ($M(^{\bullet}OH)$) and oxygen in the oxide lattice (MO_{x+1}) to the anode surface (Eq. 2.93) [23]. It can be modified by reducing, taking electrons and transferring them to oxygen [33-35]. The process can be carried out in several cycles until the properties of the anode and cathode are exhausted [36].

$$Fe \rightarrow Fe^{2+} + 2e^{-} \tag{2.92}$$

$$O_2 + 2H^{+} + 2e^{-} \rightarrow H_2O_2 \tag{2.93}$$

The process can also be carried out with external dosing of hydrogen peroxide and Fe^{2+} with reduction of Fe^{3+} formed by electrochemical means (electro-Fenton process type III). The last available option is the simultaneous electrochemical reduction of conjugated iron ions Fe^{3+} to Fe^{2+}. The reduction of dissolved oxygen to hydrogen peroxide. Both reactions occur on a single electrode (type VI electro-Fenton process) [63].

In electrochemical systems, anodes can be made of metals and metal oxides, such as boron-doped diamond (BDD), platinum (Pkt), and titanium coated electrodes (titanium substrate is coated with metal oxides, TiO_2, RuO_2 and IrO_2, etc.). The methods have a high potential for releasing O_2, but they are expensive, and the formation of radicals occurs only on the surface of the anode, destroying the surrounding impurities. It is worth noting that the system cannot be used in the construction a continuous flow of current [37]. Therefore, new materials are difficult to produce anodes, which are characterized by low cost, long service life and relatively high catalytic activity. Materials containing such as activated carbon fibre, graphite, carbon sponge and carbon or graphite felt are used for preparation. The most popular material is a carbon electrode with polytetrafluoroethylene (PTFE) due to its high diffusion efficiency [64].

Fenton reactions also use a combination of magnetite-plasma nanocomposites with plasma pulsed discharge (PDP). The plasma improved energy efficiency and to accelerate the reaction by applying ultraviolet radiation to the plasma channel. The additional energy accelerates the conversion of H_2O_2 into $HO^•$ by intensifying the ionic reaction. During a plasma discharge, oxygen molecules are excited and dissociated by high-energy electrons [65] (Eqs. 2.94–2.95):

$$O_2 + e^- \rightarrow 2O + e^- \tag{2.94}$$

$$O + O_2 + M \rightarrow O_3 + M \tag{2.95}$$

In addition, the efficiency of the process can be increased by corrosion caused by the release of an electron from the surface of the ZVI. The possibility of modifying the Fenton process favors the use of the method in many industries [33].

Heterogeneous Modifications of the Fenton Process

In the heterogeneous process, the iron needed to start the process is not introduced in the form of dissolved ions, but in the form of a solid. The concept of heterogeneous catalyst operation is shown in Figure 2.6.

The heterogeneous Fenton process consists of two healing mechanisms. First, radical oxidation in an acidic environment, secondly, coagulation and precipitation in an alkaline environment. Theoretically, radical oxidation can provide complete oxidation, but in practice it forms many different organic compounds with lower molecular weight as a result of chemical reactions. Moreover, reaction products, due to oxygen-containing functional groups, are usually more polar and have higher water solubility and lower toxicity than parent compounds. The coagulation/precipitation step increases the efficiency of purification due to the adsorption of organic contaminants on hydroxide surfaces. On the other hand, another disadvantage of the Fenton method is the significant amount of sludge during the final coagulation or neutralization reaction. In the final phase of the process, it forms simpler forms of compounds, suitable for biological purification and usually less toxic [24].

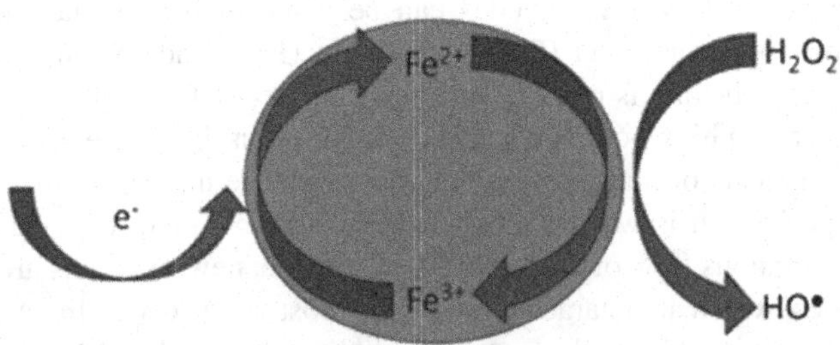

FIGURE 2.6 Heterogeneous catalyst concept.
Source: Own elaboration.

Thanks to parameter modifications, several types of processes can be distinguished, such as heterogeneous [15,38], electro-Fenton [39,40], electro-photo-Fenton, photo-Fenton [2,41], and sono-Fenton [42]. All this can be obtained using various sources of iron, such as metallic iron or other compounds containing iron in their structure. The use of heterogeneous catalysts with additional energy from UV light, electricity and ultrasound is becoming increasingly popular. An example of such a modification is the use of metallic iron (Fe^0, zero-valent iron, ZVI). ZVI can easily undergo various redox reactions, as its standard potential at 25°C, E_0 (Fe^{2+}/Fe^0) is −0.44 V. In this way, we can easily transfer it from metallic to ionic form. However, HFP has many drawbacks that make it much more difficult to introduce this industry. In HFP, the outer surface of the catalyst generates Fe^{2+} ions, the consumption of which increases with the reaction time. In addition, the reaction occurs throughout the entire liquid phase. For this reason, iron ions are deactivated and must be added in subsequent cycles of the process, without the possibility of recycling the catalyst [43].

Eq. (2.96) represents the oxidation of Fe^0 by H^+ in the absence of O_2 gives Fe^{2+}. One of the most important of the Eq. (2.97) in the Fenton process described the reaction of ZVI with H_2O_2, ending in the formation of water [44].

$$Fe^0 + 2H^+ \rightarrow Fe^{2+} + H_2 \tag{2.96}$$

$$Fe^0 + H_2O_2 + 2H^+ \rightarrow Fe^{2+} + 2H_2O \tag{2.97}$$

The $Fe^0/H_2O/O_2$ system generates H_2O_2, $HO^{\bullet}$ and other reactive oxygen species (ROS) under acidic conditions. The reactions occur through the two-electron oxidation of Fe^0. The next stage of the reaction is the Fenton process, presented in Eqs. 2.98–2.100 [44].

$$Fe^0 + O_2 + 2H^+ \rightarrow Fe^{2+} + H_2O_2 \tag{2.98}$$

$$2Fe^0 + O_2 + 4H^+ \rightarrow 2Fe^{2+} + 2H_2O \tag{2.99}$$

$$2Fe^0 + O_2 + 2H_2O \rightarrow 2Fe^{2+} + 4HO^- \tag{2.100}$$

Fe^{2+} oxidation by O_2 is relatively rapid, which affects the speed of reaction. The generation of Fe^{3+} in Fenton process causes hydrolysis and precipitation of iron oxide. In the next step of the reaction, Fe^{3+} transforms to oxides and hydroxides such as Fe_3O_4, $Fe(OH)_3$, FeOOH, as presented in Eqs. 2.101–2.103 [44].

$$6Fe^{2+} + O_2 + 6H_2O \rightarrow 2Fe_3O_4 \downarrow + 12H^+ \tag{2.101}$$

$$4Fe^{2+} + O_2 + 10H_2O \rightarrow 4Fe(OH)_3 \downarrow + 8H^+ \tag{2.102}$$

$$4Fe^{2+} + O_2 + 6H_2O \rightarrow 4FeOOH \downarrow + 8H^+ \tag{2.103}$$

Eq. (2.104) presents the reaction of recycling of Fe^{3+} at the iron surface [44].

$$2Fe^{3+} + Fe^0 \rightarrow 3Fe^{2+} \tag{2.104}$$

In the HFP, the treatment effects are not limited to initial oxidation and final coagulation. A lot of processes related to catalysts' surface occur. These include oxidation and reduction reactions, ion exchange, sorption, and passivation, presented in Figure 2.7. Figure 2.7 shows other processes, which happen on the catalysts' surface, such as absorption, adsorption, and passivation [36,45].

The HFP can occur with sources of iron other than metallic iron, *e.g.*, iron oxides and oxyhydroxides. In the HFP, the primary oxidant of the reaction is hydrogen peroxide and oxidation "boosters", *e.g.*, persulfate, microwave radiation, or a combination of methods [46]. In the mentioned process, the source of hydroxyl is the iron ions on the catalyst surface [36]. The reaction occurs in the active iron sites available on the catalyst surface, reacting with hydrogen peroxide. If the reaction fails, the hydrogen peroxide forms a complex with iron on the catalyst's surface [7,47]. The main advantage of the HFP is the low reduction of iron ions from the catalyst, reducing iron losses. The saved catalyst can be reused in subsequent cycles, so it is worth separating the catalyst after the completed process and subjecting it to recycling. Moreover, the amount of sludge, which forms after the process, could be reduced by limiting metal leaching in heterogeneous catalytic converters and little ion

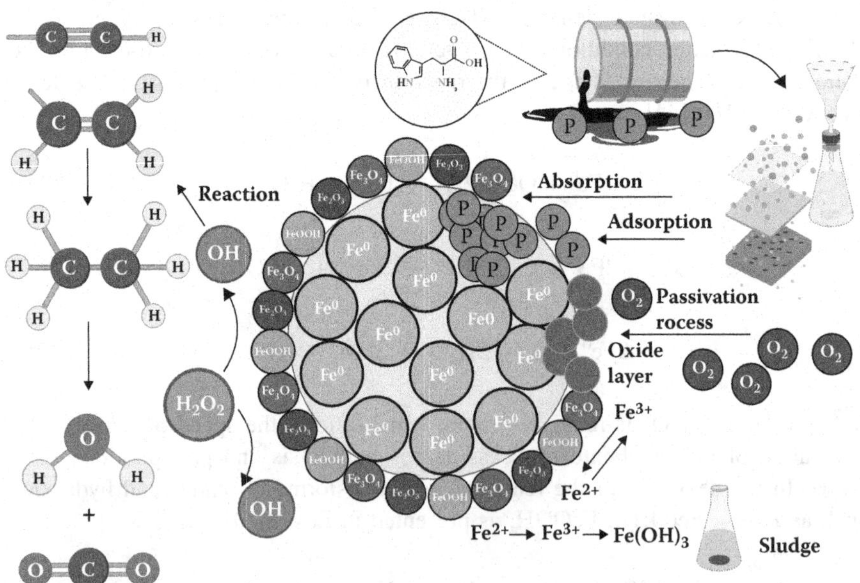

FIGURE 2.7 Fe^0 surface reaction and processes (created with BioRender.com).

Source: Own work.

elucidation from the catalyst [36,37,48]. Furthermore, during the process, the adsorption of pollutants on the catalyst, due to their porous structure surface, affects the effectiveness of reactions and can also be observed. Another advantage of the heterogeneous Fenton process is the wider pH range, in which the decomposition of pollutants is effective [49–51]. What is more, by immobilization of iron compounds on usually porous substrates such as clay, activated carbon, silica, and hydrogels, it is possible to adjust the number of iron ions on their surface and extend the life of the catalyst [36]. One of the disadvantages of this process is the difficulty in using the catalyst in the industry due to their short life cycle and low efficiency [52]. What is more, the limited surface area of the compounds, in which the reaction could occur, can generate a low mass transfer rate and slow the whole process. In the heterogeneous Fenton process, the reaction occurs in the entire liquid phase and on the catalyst surface while mixing the reactants in the reactor [46]. Selection of parameters such as pH, a dose of reagents, methods to generate the radical, duration, the type of substance to be oxidized, and additional energy to the catalyst process are one of the most important elements, which depend on the efficiency of the process [2,42,46,53].

Among the heterogeneous catalysts, not only ZVI, but also iron-based minerals, such as magnetite, hematite, goethite, schorl, and pyrite are very important [36,54]. For some of them, the properties of the semiconductor are extremely important [54], (Eqs. 2.105–2.113).

$$\text{catalyst} + h\nu \rightarrow \text{catalyst}(h_{vb}^{+} + e_{cb}^{-}) \tag{2.105}$$

$$Fe_aO_b + h\nu \rightarrow Fe_aO_b(h_{vb}^{+} + e_{cb}^{-}) \tag{2.106}$$

$$Fe_aO_b - OH + h\nu \rightarrow Fe_aO_b + HO^{\bullet} \tag{2.107}$$

$$Fe_aO_b + H_2O_2 \rightarrow Fe_aO_b - O_2H + HO^{\bullet} \tag{2.108}$$

$$Fe_aO_b - O_2H + h\nu \rightarrow Fe_cO_d + FeO^{2+} + HO^{-} \tag{2.109}$$

$$e_{cb}^{-} + O_2 \rightarrow O_2^{\bullet -} \tag{2.110}$$

$$h_{vb}^{+} + H_2O \rightarrow H^{+} + HO^{\bullet} \tag{2.111}$$

$$H_2O_2 + e_{cb}^{-} \rightarrow HO^{-} + HO^{\bullet} \tag{2.112}$$

$$h_{vb}^{+} + RH \rightarrow R^{\bullet} + H^{+} \tag{2.113}$$

RADICAL SCAVENGERS

The salinity of the solution is a factor limiting the effectiveness of the radicals. The ions that are usually present in the highest concentrations are usually sulfates, chlorides, and bicarbonates. If the process is carried out under strongly acidic conditions, decomposition of bicarbonates is expected, but their negative effect is revealed when the process is carried out at pH close to neutral conditions [55].

- Sulfates (Eqs. 2.114–2.117):

$$SO_4^{2-} + HO^{\bullet} + H^+ \rightarrow SO_4^{\bullet -} + H_2O \quad (2.114)$$

$$SO_4^{2-} + HO^{\bullet} \rightarrow SO_4^{\bullet -} + HO^- \quad (2.115)$$

$$SO_4^{\bullet -} + e \rightarrow SO_4^{2-} \quad (2.116)$$

$$HO^{\bullet} + HSO_4^- \rightarrow SO_4^{\bullet -} + H_2O \quad (2.117)$$

- Chlorides (Eqs. 2.118–2.124):

$$Cl^- + HO^{\bullet} \rightarrow Cl^{\bullet} + HO^- \quad (2.118)$$

$$Cl^{\bullet} + H_2O_2 \rightarrow Cl + H^+ + HO_2^{\bullet} \quad (2.119)$$

$$Cl^- + Cl^{\bullet} \rightarrow Cl_2^{\bullet -} \quad (2.120)$$

$$Cl_2^{\bullet -} + Cl_2^{\bullet -} \rightarrow 2Cl^- + Cl_2 \quad (2.121)$$

$$Cl_2 + H_2O \rightarrow HClO + HCl \quad (2.122)$$

$$Cl^{\bullet} + H_2O \rightarrow HOCl^{\bullet -} + H^+ \quad (2.123)$$

$$HOCl^{\bullet -} \rightarrow Cl^- + HO^{\bullet} \quad (2.124)$$

- Bicarbonates (Eq. 2.125):

$$HCO_3^- + HO^{\bullet} \rightarrow H_2O + CO_3^{\bullet -} \quad (2.125)$$

- Inter-ion reactions (Eqs. 2.126–2.130):

$$Cl^- + SO_4^{\bullet -} \rightarrow SO_4^{2-} + Cl^{\bullet} \quad (2.126)$$

$$NO_3^- + SO_4^{\bullet -} \rightarrow SO_4^{2-} + NO_3^{\bullet} \quad (2.127)$$

$$NO_2^- + SO_4^{\bullet -} \rightarrow SO_4^{2-} + NO_2^{\bullet} \quad (2.128)$$

$$CO_3^{2-} + SO_4^{\bullet-} \rightarrow SO_4^{2-} + CO_3^{\bullet-} \quad (2.129)$$

$$HCO_3^{-} + SO_4^{\bullet-} \rightarrow SO_4^{2-} + CO_3^{\bullet-} + H^{+} \quad (2.130)$$

Scientists are determined to develop methods that will be more friendly to the environment and do not generate additional pollution. The heterogeneous Fenton process is rather safe for the natural ecosystems and surrounding organisms of the trophic chain, and therefore it stands out from other methods. One of the advantages of the process is the possibility of the utilization of magnetic iron oxides. Such catalysts can be easily recovered and reused. The magnetic iron minerals can be separated by applying a magnetic field in the reactor, then separated and then redispersed in the solution for reuse [56,57]. Different research groups readily investigate the possibility of recycling the selected materials, which can be reused in the process.

REFERENCES

1. Zhang, Y. and M. Zhou, A critical review of the application of chelating agents to enable Fenton and Fenton-like reactions at high pH values. *Journal of Hazardous Materials*, 2019. **362**: p. 436–450.
2. Ebrahiem, E.E., M.N. Al-Maghrabi, and A.R. Mobarki, Removal of organic pollutants from industrial wastewater by applying photo-Fenton oxidation technology. *Arabian Journal of Chemistry*, 2017. **10**: p. S1674–S1679.
3. Munoz, M., et al., Preparation of magnetite-based catalysts and their application in heterogeneous Fenton oxidation – A review. *Applied Catalysis B: Environmental*, 2015. **176-177**: p. 249–265.
4. Kang, S.-F., C.-H. Liao, and M.-C. Chen, Pre-oxidation and coagulation of textile wastewater by the Fenton process. *Chemosphere*, 2002. **46**(6): p. 923–928.
5. Kim, T.-H., et al., Comparison of disperse and reactive dye removals by chemical coagulation and Fenton oxidation. *Journal of Hazardous Materials*, 2004. **112**(1): p. 95–103.
6. Barbusiński, K., Intensyfikacja procesu oczyszczania ścieków i stabilizacji osadów nadmiernych z wykorzystaniem odczynnika Fentona. 2004, Wydawnictwo Politechniki Śląskiej.
7. Lin, S.-S. and M.D. Gurol, Catalytic decomposition of hydrogen peroxide on iron oxide: kinetics, mechanism, and implications. *Environmental Science & Technology*, 1998. **32**(10): p. 1417–1423.
8. Vorontsov, A.V., Advancing Fenton and photo-Fenton water treatment through the catalyst design. *J Hazard Mater*, 2019. **372**: p. 103–112.
9. Rodriguez, S., A. Santos, and A. Romero, Oxidation of priority and emerging pollutants with persulfate activated by iron: effect of iron valence and particle size. *Chemical Engineering Journal*, 2017. **318**: p. 197–205.
10. Babuponnusami, A. and K. Muthukumar, A review on Fenton and improvements to the Fenton process for wastewater treatment. *Journal of Environmental Chemical Engineering*, 2014. **2**(1): p. 557–572.
11. De Laat, J. and H. Gallard, Catalytic decomposition of hydrogen peroxide by Fe(III) in homogeneous aqueous solution: mechanism and kinetic modeling. *Environmental Science & Technology*, 1999. **33**(16): p. 2726–2732.

12. Jungclaus, G., V. Avila, and R. Hites, Organic compounds in an industrial Wastewater: a case study of their environmental impact. *Environmental Science & Technology*, 1978. **12**(1): p. 88–96.
13. Brillas, E., I. Sirés, and M.A. Oturan, Electro-Fenton process and related electrochemical technologies based on Fenton's reaction chemistry. *Chemical Reviews*, 2009. **109**(12): p. 6570–6631.
14. Wu, Y., et al., Characteristics and mechanisms of kaolinite-supported zero-valent iron/H_2O_2 system for nitrobenzene degradation. *CLEAN – Soil, Air, Water*, 2017. **45**(3): p. 1600826.
15. Xu, X., et al., Cyclohexanoic acid breakdown by two-step persulfate and heterogeneous Fenton-like oxidation. *Applied Catalysis B: Environmental*, 2018. **232**: p. 429–435.
16. Sychev, A.Y. and V.G. Isak, Iron compounds and the mechanisms of the homogeneous catalysis of the activation of O_2 and H_2O_2 and of the oxidation of organic substrates. *Russian Chemical Reviews*, 1995. **64**(12): p. 1105–1129.
17. Basheer Hasan, D.U., A.A. Abdul Raman, and W.M.A. Wan Daud, Kinetic Modeling of a Heterogeneous Fenton Oxidative Treatment of Petroleum Refining Wastewater. *The Scientific World Journal*, **2014**. 2014: p. 252491.
18. Ziembowicz, S. and M. Kida, Limitations and future directions of application of the Fenton-like process in micropollutants degradation in water and wastewater treatment: a critical review. *Chemosphere*, 2022. **296**: p. 134041.
19. He, J., et al., Interfacial mechanisms of heterogeneous Fenton reactions catalyzed by iron-based materials: a review. *Journal of Environmental Sciences*, 2016. **39**: p. 97–109.
20. Bray, W.C. and M.H. Gorin, Ferryl ion, a compound of tetravalent iron. *Journal of the American Chemical Society*, 1932. **54**(5): p. 2124–2125.
21. Ortiz de la Plata, G.B., O.M. Alfano, and A.E. Cassano, Decomposition of 2-chlorophenol employing goethite as Fenton catalyst. I. Proposal of a feasible, combined reaction scheme of heterogeneous and homogeneous reactions. *Applied Catalysis B: Environmental*, 2010. **95**(1): p. 1–13.
22. Yang, X., et al., Rapid degradation of methylene blue in a novel heterogeneous Fe_3O_4 @rGO@TiO_2-catalyzed photo-Fenton system. *Sci Rep*, 2015. **5**: p. 10632.
23. Martinez-Huitle, C.A. and S. Ferro, Electrochemical oxidation of organic pollutants for the wastewater treatment: direct and indirect processes. *Chemical Society reviews*, 2006. **35**: p. 1324–1340.
24. Bogacki, J., et al., Cosmetic wastewater treatment by the ZVI/H_2O_2 process. *Environmental Technology*, 2017. **38**(20): p. 2589–2600.
25. Bogacki, J.P., et al., Cosmetic wastewater treatment using dissolved air flotation. *Archives of Environmental Protection*, 2017. **43**(2).
26. Catrinescu, C., et al., Degradation of 4-chlorophenol from wastewater through heterogeneous Fenton and photo-Fenton process, catalyzed by Al–Fe PILC. *Applied Clay Science*, 2012. **58**: p. 96–101.
27. Prousek, J., et al., Fenton- and Fenton-like AOPs for wastewater treatment: from laboratory-to-plant-scale application. *Separation Science and Technology*, 2007. **42**: p. 1505–1520.
28. Hermosilla, D., M. Cortijo, and C.P. Huang, The role of iron on the degradation and mineralization of organic compounds using conventional Fenton and photo-Fenton processes. *Chemical Engineering Journal*, 2009. **155**(3): p. 637–646.
29. Malato, S., et al., Decontamination and disinfection of water by solar photocatalysis: recent overview and trends. *Catalysis Today*, 2009. **147**(1): p. 1–59.
30. Soares, P.A., et al., Insights into real cotton-textile dyeing wastewater treatment using solar advanced oxidation processes. *Environ Sci Pollut Res Int*, 2014. **21**(2): p. 932–945.

31. Gogate, P.R. and A.B. Pandit, A review of imperative technologies for wastewater treatment I: oxidation technologies at ambient conditions. *Advances in Environmental Research*, 2004. **8**(3): p. 501–551.
32. Geng, N., et al., Insights into the novel application of Fe-MOFs in ultrasound-assisted heterogeneous Fenton system: efficiency, kinetics and mechanism. *Ultrasonics Sonochemistry*, 2021. **72**: p. 105411.
33. Fan, L., et al., Treatment of bromoamine acid wastewater using combined process of micro-electrolysis and biological aerobic filter. *Journal of hazardous materials*, 2008. **162**: p. 1204–1210.
34. Ruan, X.-C., et al., Degradation and decolorization of reactive red X-3B aqueous solution by ozone integrated with internal micro-electrolysis. *Separation and Purification Technology*, 2010. **74**(2): p. 195–201.
35. Zhou, Y., et al., Degradation of 3,3 '-iminobis-propanenitrile in aqueous solution by Fe-0/GAC micro-electrolysis system. *Chemosphere*, 2012. **90**.
36. Nidheesh, P.V., Heterogeneous Fenton catalysts for the abatement of organic pollutants from aqueous solution: a review. *RSC Advances*, 2015. **5**(51): p. 40552–40577.
37. Guo, S., G. Zhang, and J. Wang, Photo-Fenton degradation of rhodamine B using Fe2O3–Kaolin as heterogeneous catalyst: characterization, process optimization and mechanism. *Journal of Colloid and Interface Science*, 2014. **433**: p. 1–8.
38. Zárate-Guzmán, A.I., et al., Towards understanding of heterogeneous Fenton reaction using carbon-Fe catalysts coupled to in-situ H_2O_2 electro-generation as clean technology for wastewater treatment. *Chemosphere*, 2019. **224**: p. 698–706.
39. Zhang, H., et al., Pretreatment of shale gas drilling flowback fluid (SGDF) by the microscale Fe0/Persulfate/O3 process (mFe0/PS/O3). *Chemosphere*, 2017. **176**.
40. Wang, Z., et al., Removal of COD from landfill leachate by advanced Fenton process combined with electrolysis. *Separation and Purification Technology*, 2019. **208**: p. 3–11.
41. Yang, Z., et al., Enhanced Nitrobenzene reduction by zero valent iron pretreated with H_2O_2/HCl. *Chemosphere*, 2018. **197**: p. 494–501.
42. Rezaei, F. and D. Vione, Effect of pH on zero valent iron performance in heterogeneous Fenton and Fenton-like processes: a review. *Molecules*, 2018. **23**(12): p. 3127.
43. Duarte, F., F.J. Maldonado-Hódar, and L.M. Madeira, Influence of the characteristics of carbon materials on their behaviour as heterogeneous Fenton catalysts for the elimination of the azo dye Orange II from aqueous solutions. *Applied Catalysis B: Environmental*, 2011. **103**(1): p. 109–115.
44. Litter, M.I. and M. Slodowicz, An overview on heterogeneous Fenton and photoFenton reactions using zerovalent iron materials. *Journal of Advanced Oxidation Technologies*, 2017. **20**(1).
45. Campos, S., et al., Nafcillin degradation by heterogeneous electro-Fenton process using Fe, Cu and Fe/Cu nanoparticles. *Chemosphere*, 2020. **247**: p. 125813.
46. da Fonseca, F., et al., Heterogeneous Fenton process using the mineral hematite for the discolouration of a reactive dye solution. *Brazilian Journal of Chemical Engineering*, 2011. **28**: p. 605–616.
47. Xue, X., et al., Adsorption and oxidation of PCP on the surface of magnetite: kinetic experiments and spectroscopic investigations. *Applied Catalysis B: Environmental*, 2009. **89**(3): p. 432–440.
48. Nie, Y., et al., Enhanced Fenton-like degradation of refractory organic compounds by surface complex formation of $LaFeO_3$ and H_2O_2. *Journal of Hazardous Materials*, 2015. **294**: p. 195–200.
49. Kušić, H., N. Koprivanac, and I. Selanec, Fe-exchanged zeolite as the effective heterogeneous Fenton-type catalyst for the organic pollutant minimization: UV irradiation assistance. *Chemosphere*, 2006. **65**(1): p. 65–73.

50. Kasiri, M.B., H. Aleboyeh, and A. Aleboyeh, Degradation of Acid Blue 74 using Fe-ZSM5 zeolite as a heterogeneous photo-Fenton catalyst. *Applied Catalysis B: Environmental*, 2008. **84**(1): p. 9–15.
51. Kusic, H., N. Koprivanac, and A. Loncaric Bozic, Degradation of organic pollutants by zeolite assisted AOPs in aqueous phase: heterogeneous Fenton type processes, in *13th International Conference on Advanced Oxidation Technologies for Treatment of Water, Air and Soil*. 2007: Niagara Falls, NY, USA.
52. Zhang, G., S. Wang, and F. Yang, Efficient adsorption and combined heterogeneous/homogeneous Fenton oxidation of amaranth using supported nano-FeOOH as cathodic catalysts. *The Journal of Physical Chemistry C*, 2012. **116**(5): p. 3623–3634.
53. Ribeiro, A.R., et al., An overview on the advanced oxidation processes applied for the treatment of water pollutants defined in the recently launched Directive 2013/39/EU. *Environment International*, 2015. **75**: p. 33–51.
54. Thomas, N., D.D. Dionysiou, and S.C. Pillai, Heterogeneous Fenton catalysts: a review of recent advances. *Journal of Hazardous Materials*, 2021. **404**: p. 124082.
55. Ahmed, N., et al., A review on the degradation of pollutants by Fenton-like systems based on zero-valent iron and persulfate: effects of reduction potentials, pH, and anions occurring in waste waters. *Molecules*, 2021. **26**(15): p. 4584.
56. Li, X.Z., et al., Photocatalytic oxidation using a new catalyst--TiO_2 microsphere--for water and wastewater treatment. *Environ Sci Technol*, 2003. **37**(17): p. 3989–3994.
57. Lai , B.-H. , C.-C. Tak, and D.-H. Chen ,Modyfikacja powierzchni nanocząstek tlenku żelaza za pomocą poliargininy jako wysoce dodatnio naładowanego nanoadsorbentu magnetycznego do szybkiego i skutecznego odzyskiwania kwaśnych białek . *Biochemia procesu*, 2012. **47**(5): p. 799–805.
58. Ovalle, R., "A history of the Fenton reactions (Fenton chemistry for beginners)." In R. Ahmad, (ed.), Reactive Oxygen Species, Biochemistry, 2022. IntechOpen p. 7.
59. Hussain, S., E. Aneggi, and D. Goi, Catalytic activity of metals in heterogeneous Fenton-like oxidation of wastewater contaminants: a review. *Environmental Chemistry Letters*, 2021. **19**: p. 2405–2424.
60. Hussain, S., E. Aneggi, and D. Goi, Catalytic activity of metals in heterogeneous Fenton-like oxidation of wastewater contaminants: a review. *Environmental Chemistry Letters,* 2021. **19**: p. 2405–2424.
61. Lee, J., B.K. Singh, M.A. Hafeez, K. Oh, and W. Um, Comparative study of PMS oxidation with Fenton oxidation as an advanced oxidation process for Co-EDTA decomplexation. *Chemosphere*, 2022. **300**: p. 134494.
62. Ziembowicz, S., and M. Kida, Limitations and future directions of application of the Fenton-like process in micropollutants degradation in water and wastewater treatment: a critical review. *Chemosphere*, 2022. **296**: p. 134041.
63. Zhang, M., H. Dong, L. Zhao, D. Wang, and D. Meng, A review on Fenton process for organic wastewater treatment based on optimization perspective. *Science of The Total Environment*, 2019. **670**: p. 110–121.
64. Divyapriya, G. and P.V. Nidheesh, Importance of graphene in the electro-Fenton process. *ACS Omega*, 2020. **5**: p. 4725–4732.
65. Guo, H., Z. Li, Y. Zhang, N. Jiang, H. Wang, and J. Li, Degradation of chloramphenicol by pulsed discharge plasma with heterogeneous Fenton process using Fe_3O_4 nanocomposites. *Separation and Purification Technology*, 2020. **253**: p. 117540.

3 Catalysts

The catalyst has a significant role in the heterogeneous Fenton process the type as well as size determine the specific surface area, where the reaction takes place, the number of iron ions, the rate of leaching metallic iron and iron compounds, and the possibility of reusing the catalyst. Compounds such as metal compounds, and materials with carbon in their structure, can be used as catalysts [1]. The most popular ones are iron-based materials due to their highly active and promote decomposition compounds. The type of catalyst can also influence the radical generation, which is dependent on the oxygen deficiency on their surface due to charge exchange and cation deficiency in the compounds, which can react in the process. The sites of oxygen deficiency constitute the active specific surface, where the decomposition of H_2O_2 takes place and the process accelerates. Moreover, iron catalysts are cheap, low toxic, and easy to gather and reuse in the process. Additionally, the catalysts could be used in different configurations, modifications, and structural forms as catalysts do decompose huge types of pollutants in various conditions and parameters of the process (see Table 3.1). One of the significant problems is the deactivation of the catalyst and the leaching of material. However, in the heterogeneous Fenton process, the leaching of Fe^{2+} from the catalyst depends on compounds and the kind of support used in the process. Also, the catalyst can be separated and reused by recycling the material. Moreover, the catalysts have a porous structure, which enables the adsorption of pollutants on their surface. The assessments and properties of iron compounds such as crystallinity [2], surface area [3], and activity are related to the leaching of a metal compound in an acidic environment, which affects the effectiveness of catalyst in heterogeneous Fenton process [4]. Besides, it is also essential to choose the correct dose of the catalyst because too little can be ineffective due to the small surface area in which the process takes place [5]. Zhang *et al.* [6] noted that applying the process with additional catalysts is more effective. They confirmed it by the results of K_{obs} values (observed rate constant), which were 1.68, 0.69, and 0.28 1/h in processes ZVI/ethylenediaminetetraacetic acid/air, ZVI/Air, and ZVI/ethylenediamine-N,N'-disuccinic acid/air, respectively. The catalysts can be modified and used in various combinations by altering their structure, contributing to the increased adsorption and degradation of organic pollutants [7]. To increase the magnetite adsorption capacity and introduce other transition metals into their structure scientists used chelating agents, pillar clays, activated carbon, especially carbon-containing materials [8], carbon xerogels [9], carbon nanotubes, foam, and graphite; and alumina support [10,11]. Catalysts differ in their physical, chemical, and reactive properties. Scientists divided them into five categories: zero-value iron, iron minerals, iron oxides (hydrogen), multi-metals, iron-based materials, and supported iron-based materials [12].

DOI: 10.1201/9781003364085-3

TABLE 3.1
The Types of the Catalysts and Their Efficiency in the Heterogenous Fenton Process in Various Parameters and Conditions

Ref.	Brief Description of the Experiment	Catalysts and Other Process Reagents	Industry/Type of Pollution	Pollution Parameters	Analysis of Catalysts	Reagent Doses and Process Parameters	Optimal Conditions	Process Efficiency (Decrease in Parameters)
[104]	Decolorization and mineralization of textile wastewater in a heterogeneous Fenton like system with/without ultrasound and ultraviolet	ZVI (Zero Valent Iron)	Reactive Black 5 (RB5) EDTA	RB5 = 100 mg/L EDTA = 0,4 mM	HPLC, IC, UV-vis, pH meter, TOC analyzer	pH = 2.0, 3.0, 4.0, 7.0 Time = 0-180 min EDTA = 0.1, 0.4, 1.0, 5.0 mM ZVI = 1, 5, 25, 50 g/L	pH = 7.0 Time = 180 min EDTA = 0.1 – 1.0 mM ZVI = 25 g/L	RB5 = 100%, EDTA = 96.5% TOC = 68.6% COD = 92.2%
[6]	Removal of 2,4-dichlorophenol from contaminated soil by a heterogeneous ZVI/EDTA/Air Fenton-like system	ZVI/EDTA/Air ZVI/Air ZVI/EDDS/Air	2,4-dichlorophenol from contaminated soil	20 g	HPLC, IC, XCT ion trap, ART-FTIR, atomic absorbing spectrometer, pH meter, dissolved oxygen meter	Time = 0, 15, 30, 45, 60, 90, 120, 180 min EDTA = 0.2, 0.4, 0.75, 1.6, 2.0 mmol/L ZVI = 2.5, 7.5, 12.5 g/L Air = 0.5, 1.0, 2.0 L/min	Time = 45 min/L EDTA = 0.4 mmol/L ZVI = 7.5 g/L Air = 1.0 L/min	96% 2,4-DCP = 96% degradation
[5]	Heterogeneous Fenton Process using the mineral hematite for discoloration of a reactive dye solution	Hematite	Drimarene Red X-6BN (C.I. Reactive Red 243)	100 mg/L	UV-VIS, atomic absorption spectrometry, colorimetric method, BET	pH = 2.5, 3.5, 5.0 Time = 0, 20, 40, 60, 80, 100, 120 min H_2O_2 = 0, 200, 500, 800 mg/L Hematite = 1, 5, 10, 20 g/L Temp = 25, 35, 45, 55°C	pH = 2.5 Time = 120 min H_2O_2 = 800 mg/L Hematite = 20 g/L Temp = 25	Reactive Red 243 = 99% degradation

[105]	Heterogeneous Fenton-type processes for the degradation of organic dye pollutants in water — The application of zeolite-assisted AOPs	$FeZSM_5/H_2O_2$, $UV/FeZSM_5/H_2O_2$, UV, $UV/FeZSM_5$, UV/Fe^{3+}, UV/H_2O_2,	RB137 (C.I. Reactive Blue 137	20 mg/L in 0.5 L	FTIR spectroframs, IC-MSP analyses	Time = 0, 2, 5, 10, 20, 30, 40, 50, 60 min pH = 2.0–8.0 ZSM_5 = 0.745 i 1.49 g/L Fe = 0, 5, 1,0 mM H_2O_2 = 10 mM	$UV/FeZSM_5/H_2O_2$ $FeZSM_5$ = 1.49 g/L H_2O_2 = 10 mM pH = 6	TOC = 81.1% RB137 = 100%
[95]	Graphene-modified iron sludge derived from an efficient heterogeneous Fenton catalyst for degradation of organic pollutant	Iron sludge with a low amount (0–2 wt%) of graphene	Rhodamine B, 23 acid red G, metronidazole	10 mg/L	XRD, FESEM, TEM, Raman, XPS, BET, FTIR UV-vis	pH = 3.03, 6.01, 9.44 Fe-G = 0.2, 0.5, 1.0 g/L H_2O_2 = 0, 1, 5, 10 mmol/L	pH = 3.03 Fe-G = 1 g/L H_2O_2 = 10 mmol/L	Rhodamine B = 99.0% 23 acid red G = 98.5% Metronidazole = 91.8%
[96]	Heterogeneous Electro-Fenton Using Modified Iron-carbon as Catalyst for 2,4-dichlorophenol	Fe-C catalyst	2,4-dichlorophenol (2,4-DCP)	2,4-DCP = 120 mg/L TOC = 81 I 277 mg/L	FE-SEM, XPS, XRD, HPLC, TOC analyzer, ion chromatography	Time = 0–120 min Current = 50–200 mA Fe-C = 2–6 g/L pH = 3.0, 5.0, 6.0, 7.0, 9.0	pH = 2.72 Fe-C = 6 g/L Current = 100 mA Time = 120	2,4-DCP = 100% degradation TOC = 47.3%
[136]	Azo-dye Orange II degradation by heterogeneous Fenton-like reaction	Fe-supports: an activated carbon and a carbon aerogel.	Azo dye Orange II (OII)	0.2 L of a 0.1 mM	SEM, HRTEM, XRD, XPS, BET, FTIR, UV-Vis	H_2O_2 = 3–48 mM M-Fe/H-Fe = 0.15, 0.20, 0.30 g/L pH = 2.0–4.0 T = 10–70°C	T = 30°C, pH = 3.0 H_2O_2 = 6 mM M-Fe/H-Fe = 91.5 mg/L	For M-Fe OII = 94,6%

[144]	Heterogeneous photo-Fenton processes using zero-valent iron microspheres for the treatment of wastewater contaminated with 1,4-dioxane	Zero Valent Iron (Fe^0) microsphere	1,4-Dioxane	COD = 450 mg/L;
				alkalinity = 900 mgCaCO3/L;
				conductivity = 810 lS/cm;
				1,4 dioxane = 248 mg/L
					SEM, EDS, BASF, TOC/TN, GC	H_2O_2/COD = 1.625, 2.125, 2.625 4.00
						pH = 2.8; 7.0, 8.2
						H_2O_2/Fe^0 = 1, 15, 30, 60
							H_2O_2/COD = 2.625
							pH=2.8
							H_2O_2/Fe^0 = 60
								TOC = 94%
								ChZT = 99% and 60%

(Continued)

TABLE 3.1 (Continued)
The Types of the Catalysts and Their Efficiency in the Heterogenous Fenton Process in Various Parameters and Conditions

Ref.	Brief Description of the Experiment	Catalysts and Other Process Reagents	Industry/Type of Pollution	Pollution Parameters	Analysis of Catalysts	Reagent Doses and Process Parameters	Optimal Conditions	Process Efficiency (Decrease in Parameters)
[28]	Continuous-flow heterogeneous electro-Fenton reactor for Tartrazine degradation	Iron-carbon granules	Tartrazine	40, 60, 80, 100 mg /L,	TOC, spectrophotometer	Fe-C = 86, 120, 160, 195 g Time = 0–30 min pH = 3.0, 5.0, 7.0, 9.0 H_2O_2 = 0.05, 3.0, 20.0 mL/min The flow of pollutant = 10, 15, 20, 30 mL/min	The flow of pollutant = 20 mL/min pH = 3.0 Fe-C = 160 g H_2O_2 = 20 mg/L	Tartazine = 80% degradation TOC = 30%
[86]	Fabrication of magnetic carbon composites from peanut shells and its application as a heterogeneous Fenton catalyst in the removal of methylene blue	Magnetic carbon composites from peanut shells (PMC-1, PMC-2, PMC-3 + $K_2S_2O_8$), $FeCl_2$ + $K_2S_2O_8$, Fe_3O_4 + $K_2S_2O_8$	Methylene blue	5, 10, 20, 25, 35 mg/L	LC-MS, MÖssbauer spectroscopy, XPS, SEM, XRD, the Brunauer-Emmett-Teller surface area method, BET, VSM analyses,	$K_2S_2O_8$ = 0.1, 0.25, 0.5, 0.8, 1.0 g/L pH = 3.0, 4.0, 5.0, 6.0, 8.0 Catalyst = 10, 20, 40, 60, 80 mg	$K_2S_2O_8$ = 1 g/L pH = 3.0 Catalyst = 60 mg	PMC-2 + K2S2O8 degradation 90% dye
[87]	Novel magnetic porous carbon spheres derived from the chelating resin as a heterogeneous Fenton catalyst for the removal of methylene blue from aqueous solution	Porous magnetic carbon spheres (MCS) pore-size-distribution curves the MCS-500, MCS-800, MCS-1100	Methylene blue	20 ml, 40 mg/L	X-ray diffraction, MÖssbauer spectroscopy, Raman spectroscopy, XRS, Brunauer–Emmett–Teller surface area method, SEM, VSM, LC-MS,	NH_2OH = 2.5, 5.0, 10.0 mmol/L, H_2O_2 = 1.5, 2.5, 5.0 mmol pH = 3.76, 5.14, 7.08 Catalyst = 10, 20, 30 g	H_2O_2= 5.0 mmol/L NH_2OH = 2.5 mmol/L pH = 3.76 Catalyst = 20 g	100% degradation dye (40 mg/L MB)

[91]	Niobium substituted magnetite as a strong heterogeneous Fenton the catalyst for wastewater treatment	$Fe_{3-x}Nb_xO_4$ (x = 0.0, 0.022, 0.049, 0.099, 0.19)	Methylene blue	12.5, 50, 75 i 100 mg/L	XRD, BET surface area, TEM, VSM, XPS, chemical experiments, SCA	$Fe_{3-x}Nb_xO_4$ (x = 0.0, 0.022, 0.049, 0.099, and 0.19), Time = 30–180 min, pH = 4.0, 7.0, 8.0, 10.0	$Fe_{3-x}Nb_xO_4$ series, x = 0.19 Time = 180 min pH = 7.0, 10.0	Degradation = 100% degradation dye
[145]	Treatment of textile wastewater by heterogeneous Fenton process using a new composite Fe_2O_3/carbon	Composite Fe_2O_3/carbon	Textile wastewater	Dye = 100–300 g/L COD = 1000 mg/L BOD_5 = 300 mg/L	BET, SEM/EDAX, PZC	Catalyst = 100–300 g/L H_2O_2 = 500–1000 mg/L pH = 3.0–9.0	Catalyst = 300 g/L H_2O_2 = 500 mg/L pH = 3.0	COD = 71%
[146]	Industrial wastewater treatment using hydrodynamic cavitation and heterogeneous Fenton processing	Zero valent iron (ZVI)	Industrial wastewater	ChZT = 42,000 mg/L TOC = 14,000 mg/L.	TOC analyzer	Hydrocavitator pressure = 500, 1000, 1500 psi	Hydrocavitator pressure = 1500 psi	TOC = 60%
[147]	Three-dimensional heterogeneous electro-Fenton oxidation of biologically pretreated coal gasification wastewater using sludge derived carbon as catalytic particle electrodes and catalyst	CPEs (SAC-Fe, SAC, CAC, Fe_3O_4 MNPs) were filled into real pure Fe_3O_4 magnetic nanoparticles (MNPs);	Biologically pretreated coal gasification wastewater	ChZT = 173.3 ± 14.1 mg/L TOC = 57.6 ± 4.2 mg/L BZT_5 = 14.6 ± 4.1 mg/L Total fenols = 48.3 ± 3.3 mg/L	BET, zeta potential analyzer, plasma atomic absorption spectrophotometry, SEM, XRF, XRD, FTIR, TOC analyzer, GC-MS, ESR	pH = 2.0, 3.0, 5.0, 7.0 Catalyst = 2.5, 5.0, 7.5, 10.0 g/L Current density = 10, 15, 20, 25 mA/cm^2	pH = 3.0 Catalyst = 2.5–10 g/L Current dentisity = 25 mA/cm^2	Total fenols = 78,1%, TOC = 93.5%, COD = 78.1%

(*Continued*)

TABLE 3.1 (Continued)
The Types of the Catalysts and Their Efficiency in the Heterogenous Fenton Process in Various Parameters and Conditions

Ref.	Brief Description of the Experiment	Catalysts and Other Process Reagents	Industry/Type of Pollution	Pollution Parameters	Analysis of Catalysts	Reagent Doses and Process Parameters	Optimal Conditions	Process Efficiency (Decrease in Parameters)
[148]	Heterogeneous photo-Fenton oxidation of acid orange II over iron-sewage sludge-derived carbon under visible irradiation	sewage sludge derived carbon (SC) employed as the support of iron oxide-containing catalyst (FeSC)	Acid orange II (AOII)	100 mg/L	inorganic elemental composition, XRD, SEM and TGA-FTIR, UV-vis	Catalyst = 0.5, 1.0, 1.5, 2, g/L; pH = 3.0, 4.0, 5.0	Catalyst = 1.5 g/L pH = 4.0	Decolorization = 100%
[149]	Treatment of textile wastewater by homogeneous and heterogeneous Fenton oxidation processes	Mesoporous Activated Carbon (MAC) with ferrite iron	Textile wastewater	COD = 564 mg/L BOD = 120 mg/L TOC = 144 mg/L Total solids = 7,508 mg/L Oxidation-reduction potencial = +525 mV Total dissolved solids = 7,148 mg/L Total suspended solids = 360 mg/L Chloride = 1749 mg/L Sulfate = 1,605 mg/L Ammonia = 8.4 mg/L TKN=16.2 mg/L Sulfacant =	FT-IR, UV–Visible spectroscopy, cyclic voltammetry, XRD, Diffuse Reflectance Spectroscopy	pH = 2.5, 3.5, 4.5, 5.5, 7.0 H_2O_2 = 3, 6, 9, 12,15 mM/L $FeSO_4$ = 0.4, 0.8, 1.0, 1.4,1.8 mM/L Catalys t = 2, 5, 7,5, 10, 15, 20, 25 g/L	pH = 3.5 H_2O_2 = 9–16 mM/L Catalyst = 10 g/L $FeSO_4$ = 0,8-1,8 mM/L	COD = 51 mg/L, BOD = 20 mg/L, TOC = 25 mg/L, Total solids = 7028 mg/L, Oxidation reduction potencial = +525 mV, Total dissolved solids = 6752 mg/L, Total suspended solids = 276 mg/L, chloride = 1582 mg/L, sulfate = 1281 mg/L, ammonia = 0

				9.71 mg/L Total hardness 400 Ca = 60 mg/L Volatile dissolved solids = 1,064 mg/L				mg/L, TKN = 6,0 mg/L, surfactant = 1,53 mg/L, total hardness 260, Ca = 46 mg/L, volatile dissolved solids = 0,347 mg/L
[150]	A Comparative Study of Homogeneous and Heterogeneous Photo-Fenton Process for Textile Wastewater Treatment	Copper Modified Iron Oxide	Textil wastewater	COD = 780 mg/L BOD = 198 mg/L TSS = 48 mg/L TDS = 2,603 mg/L Alkalinity = 700 mg/L	UV–Vis spectrophotometer	pH = 2.0, 3.0, 4.0, 5.0 Catalyst = 5, 10, 15, 20 mg/L H_2O_2 = 50, 100, 150, 200 mM UV power = 0, 8, 16, 24 W	pH = 3.0 Catalyst = 10 mg/L H_2O_2 = 150 mM UV = 16 W	COD removal of 62% and Dye removal of 85%
[151]	Application of heterogeneous photo-Fenton process for the mineralization of imidacloprid containing wastewater	Waste iron oxide	The imidacloprid-containing wastewater	0.58 g/L	XRD, TOC analyzer,	Time = 0–360 min pH = 3.0, 3.5, 4.0, 5.0 H_2O_2 = 26.3, 52.5, 105.0, 125.5, 157.5 mM, Catalyst=1.0, 2.0, 5.0, 10.0 g/L^{-1}	Time = 6 h pH = 3.5 H_2O_2 = 105.0 mM Catalyst = from 1.0 to 5.0 g/L	97.7% = TOC
[152]	Photo-assisted heterogeneous Fenton-like process for treatment of PNP wastewater	$Fe-Ce/Al_2O_3$	The sodium salt of p-nitrophenol (PNP)	200 mg/L	SEM, BET	H_2O_2 = 50, 75, 100, 150, 200 mmol/L Catalyst = 30, 50, 100, 200, 300 g/L, pH = 3.0, 4.0, 5.0, 6.0, 7.0, 8.0 T = 15, 25, 35, 45, 55, 65, 75°C Time = 15, 30, 45, 60, 75, 90 min.	H_2O_2 = 1.15 mmol Catalyst = 10 g Fe-Ce/Al2O3, pH = 4 Temeprature = 25°C, Time = 120 min	PNP = 98.7% COD = 72.9%

(Continued)

TABLE 3.1 (Continued)

The Types of the Catalysts and Their Efficiency in the Heterogenous Fenton Process in Various Parameters and Conditions

Ref.	Brief Description of the Experiment	Catalysts and Other Process Reagents	Industry/Type of Pollution	Pollution Parameters	Analysis of Catalysts	Reagent Doses and Process Parameters	Optimal Conditions	Process Efficiency (Decrease in Parameters)
[153]	Treatment of synthetic textile wastewater by homogeneous and heterogeneous photo-Fenton oxidation	Zero vailent iron (ZVI)	Remazol Red RR	100 mg/L	UV–vis spectrophotometer, lab kits LCK, LCI 500, LASA 100 spectrophotometer,	Catalyst = 0.25, 1, 3, 5 mM H_2O_2 = 1, 3, 10, 12, 15, 29 mM Time = 0–8 h	Catalyst = 0.25 mM H_2O_2 = 12–20 mM Time = 8 h	COD = 93%
[154]	Treatment of Actual Chemical Wastewater by a Heterogeneous Fenton Process Using Natural Pyrite	Pyrite FeS_2, Zero vailent iron (ZVI), Magnetite	Chemical wastewater	COD = 7500–8000 mg/L BOD_5/COD = 0.1 TOC = 2000 mg/L	excitation-emission matrix (EEM) analysis, BET, three-dimensional fluorescence	Catalyst = 5, 10, 20, 50 g/L, H_2O_2 = 1, 2, 5, 10, 50, 100 mmol/L pH = 1.8, 3.7 Time = 0–120 min	Catalyst = 10 g/L H_2O_2 = 50 mmol/L pH = 1.8 does 7. Time = 120 min	COD reduction of 36%
[155]	In-situ heterogeneous via photocatalysis-Fenton reaction with enriched photocatalytic performance for removal of complex wastewater	Fe-g-C_3N_4, M-Fe-g-C_3N_4 (Fe-doped g-C_3N_4)	Coking wastewater	COD = 64.6 mg/L TOC = 25.3 mg/L	XRD, FE-SEM, FTIR, XPS, UV-vis, TEM, ZETA potential and nanoparticle size analyzer, LCMS, spectrometer,three-electrode quartz cell, atomic absorption spectrophotometer, HPLC, TOC analyzer	Catalyst = 1.5, 3.0 g/L H_2O_2 = 8, 16, 20 mM	Catalyst = 1.5 g/L H_2O_2 = 8 mM	COD = 57.2% TOC = 76.1%

[108]	Heterogeneous Fenton process using steel industry wastes for methyl orange degradation	Steel industry wast (diffraction as magnetite Fe_3O_4, hematite Fe_2O_3, and wuestite).	Orange II	20 mg/L	XRD, atomic absorption spectrometer, UV-vis spectrophotometer	pH = 2.0, 3.0, 4.0, 5.0 H_2O_2 = 12, 24, 48, 68 mM Catalyst = 50, 100, 200, 250 mg/L Time = 15, 30, 45, 60, 90, 120 min	pH = 2.0 H_2O_2 = 34 mM Catalyst = 200 mg/L Time = 120 min	Orange II = 98% degradation
[99]	Hydroxamic acid mediated heterogeneous Fenton-like catalysts for the efficient removal of textile wastewater	iIron-Hpo ligand catalyst supported on granular activated carbon (GAC) HpOFe-GAC, HqFe-GAC and FeCit-GAC	Acid Red 88	30, 50, 70, 90 μM	UV–vis absorption spectra, atomic absorption spectroscopy (AAS), SEM, EDS,	Catalyst = 2.5, 5, 7.5, and 10 g/L H_2O_2 = 40 mM pH = 3.0, 5.0, 7.0, 9.0, 11.0 Temperature = 30, 40, 50, 60°C Time = 10, 20, 30, 40, 50, 60 min	Catalyst = 5 g/L, H_2O_2 = 40 mM pH = 7.0 Temperature = 50°C, Time = 60 min	Acid Red = 99.8% degradation
[102]	Removing organic contaminants with bifunctional iron-modified rectorite as efficient adsorbent and visible light photo-Fenton catalyst	Iron-modified rectorite (FeR)	Rhodamine B	80 mM	TOC analyzer, atomic absorption spectroscopy (AAS), UV–vis spectrophotometer, FTIR–ATR, ESR, XRD, XPS, SEM, BET, HRTEM	pH = 2.0, 4.0, 6.0, 8.0, 10.0 H_2O_2 = 2, 4, 6, 8, 10 mM FeR-B = 0.2, 0.4 g/L Time = 2, 4, 6 h	pH = 4.5 H_2O_2 = 6,0 mM FeR = 0.4 g/L Time = 6 h	Rhodamine B = 99% degradation
[156]	Influence of Operational Parameters in the Heterogeneous PhotoFenton Discoloration of Wastewaters in the Presence of an IronPillared Clay	Iron-pillared Tunisian clay (Fe-PILC)	Congo Red	230 mg/L	XRF, XRD, BET, FTIR methods, SEM, IR spectroscopy, UV-vis spectroscopy	Catalyst = 0.2, 0.4, 0.6, 0.8, 1.0, 1.2 g/L H_2O_2 = 100, 200, 300, 400 mg/L pH = 2.0, 4.0, 6.0, 8.0 Time = 10, 20, 30, 40, 50, 60 min	Catalyst = 0.3 g/L, H_2O_2 = 200 mg/L pH = 3.0 Time = 60 min	Congo Red = 100% degradation

(*Continued*)

TABLE 3.1 (Continued)
The Types of the Catalysts and Their Efficiency in the Heterogenous Fenton Process in Various Parameters and Conditions

Ref.	Brief Description of the Experiment	Catalysts and Other Process Reagents	Industry/Type of Pollution	Pollution Parameters	Analysis of Catalysts	Reagent Doses and Process Parameters	Optimal Conditions	Process Efficiency (Decrease in Parameters)
[157]	Degradation of Methylene Blue by Heterogeneous Fenton Reaction Using Titanomagnetite at Neutral pH Values	Titanomagnetit-e, magnetite	Methylene Blue	100 mg/L	SEM, XRD, Mossbauer spectroscopy, BET, gas sorption analyzer, FTIR, UV-vis	Titanomgnetite = 0.5, 1.0, 2.0, 3.0 g/L H_2O_2 = 0.075, 0.15, 0.30, 0.45 mol/L	Titanomagnetite = 1 g/L, H_2O_2 = 0.3 mol/L	Methylene blue = 100% degradtion
[110]	Degradation of mixed dye via heterogeneous Fenton process: Studies of calcination, toxicity evaluation, and kinetics	Iron was impregnated in black soil (Fe-BS)	The mixture of Azure B and Congo red	200 mL	FTIR, XRD, UV-Vis spectroscopy	H_2O_2 = 0.1, 0.2, 0.3, 0.4, 0.5 ml Catalyst = 2, 4, 6, 8% iron loaded into the black soil (0.1, 0.2, 0.3, 0.4 g) pH = 2.0, 3.0, 4.0, 5.0	H_2O_2 = 0.4 mL Catalyst = 0.3 g dose of 8% Fe Temperature = 150°C pH = 5.0	Degradation = 99,82%
[98]	Efficient degradation of tetracycline by a heterogeneous electro-Fenton process using Cu-doped Fe@ Fe_2O_3	Cu-doped Fe@ Fe_2O_3	Tetracycline	20 mg/L	SEM, TEM, XRD, XPS, EDX, STEM, UV–visible spectrophotometer, HPLC-MS, TOC analyzer, ICP-OES, high-performance liquid-chromatography	Time = 20, 40, 60, 80, 100, 120 min pH = 3.0, 5.0, 7.0 Cu-doped Cu/Fe mass ratio = 5, 10, 30, 50, 60%	Time = 120 min pH = 3.0 Ratio = 50 wt% Cu and 0.1 L/min aeration rate	98.1% degradation (2 h) mineralization efficiency of TC 89.8% after 6 h

[158]	Treatment of pharmaceutical wastewater by heterogeneous Fenton process: an innovative approach	ZnO nanoparticles	Pharmaceutical wastewater	100 mg/L norfloxacin	TEM, UV-vis spectrophotometer, SPR, XRD, SEM, UV-vis measurements, LC-MS	Time = 20, 40, 60, 80, 100, 120 min H_2O_2 = 50, 100, 200, 300, 400, 500 mg/L ZnO = 0, 10, 20, 30, 40, 50 mg/L pH = 2.0, 4.0, 6.0, 8.0, 10.0, 12.0	Time = 120 min H_2O_2 = 300 mg/L ZnO = 30 mg/L pH = 10.0	Norfloxacin = 98% degradation
[159]	High-efficiency heterogeneous Fenton-like catalyst biochar modified $CuFeO_2$ for the degradation of tetracycline	Biochar modified $CuFeO_2$	Tetracycline	20 mg/L	XRD, SEM, EDX, SAED, HRTEM, FTIR, XPS, BET, HPLC	H_2O_2 = 5, 10, 20, 50, 80, 100 mM Catalyst = 50, 100, 200, 500, 800, 1000 mg/L pH = 3.0, 5.0, 7.0, 9.0, 11.0	H_2O_2 = 50 mM Catalyst = 500 mg/L pH=5.0	Tetracycline = 89.12% degradation
[160]	Remediation of organic arsenic contaminants with heterogeneous Fenton process	SiO_2-coated nano zero-valent iron	p-Arsanilic acid (p-ASA)	10 mg/L	UPLC, GC-MS, ICP-MS, LiquiTOC trace, EPR, XRD, XPS	pH = 2.0, 3.0, 4.0, 5.0, 6.0, 7.0 Time = 10, 20, 30, 40, 50, 60 min	pH = 3.0 Time = 60 min	99,6% degradation of p-ASA
[161]	Application of Mineral Iron-Based Natural Catalysts in Electro-Fenton Process: A Comparative Study	Four natural catalysts, namely ilmenite ($FeTiO_3$), pyrite (FeS_2), chromite ($FeCr_2O_4$), and chalcopyrite ($CuFeS_2$)	Cefazolin antibiotic (CFZ)	0.2 mM	XRD, RAMAN, HEF	I = 50, 100, 200, 400, 500 mA, Catalyst = 0.5, 1, 2 g/L	I = 200 mA Catalyst = 1 g/L	Degradation = 100%
[162]	Magnetite and Hematite in Advanced Oxidation	$Fe_2O_3/Fe^0/H_2O_2$, $Fe_3O_4/Fe^0/H_2O_2$, light/Fe_2O_3/	Cosmetic wastewater	TOC = 146.4 mg/L COD = 819 mg/L BOD = 102 mg/L	TOC analyzer, HS-SPME-GC-MS, mass spectrometer, ChromaTOF software	H_2O_2/COD = 0,5:1, 1:1, 2:1, 4:1 Time = 15, 30, 60, 120 min	H_2O_2/COD = 0,5:1, 1:1, 2:1, 4:1 Time = 120 min Catalyst = Fe^0/Fe_3O_4	COD = 74.5 BOD_5 = 12 mg/L TOC = 69% removal

(Continued)

TABLE 3.1 (Continued)

The Types of the Catalysts and Their Efficiency in the Heterogenous Fenton Process in Various Parameters and Conditions

Ref.	Brief Description of the Experiment	Catalysts and Other Process Reagents	Industry/Type of Pollution	Pollution Parameters	Analysis of Catalysts	Reagent Doses and Process Parameters	Optimal Conditions	Process Efficiency (Decrease in Parameters)
	Processes Application for Cosmetic Wastewater Treatment	Fe0/ H_2O_2, light/Fe_3O_4/ Fe^0/ H_2O_2				Catalyst = Fe_2O_3/Fe^0/ H_2O_2, Fe_3O_4/Fe^0/ H_2O_2, light/Fe_2O_3/ Fe^0/ H_2O_2, or light/ Fe_3O_4/Fe^0/ H_2O_2 = 25/75, 50/50, 75/25, 0/100, 100/0 mg/ L with and without light	= 75/25 mg/L with light	
[45]	Magnetite, Hematite, and Zero-Valent Iron as Co-Catalysts in Advanced Oxidation Processes Application for Cosmetic Wastewater Treatment	Fe_3O_4/ Fe_2O_3/Fe^0	Cosmetic wastewater	TOC = 306.3 mg/L	TOC analyzer, HS-SPME–GC–MS	H_2O_2/COD = 1:1, 2:1, Time = 15, 30, 60, 120 min Catalyst: UV/ H_2O_2/ Fe_3O_4/Fe_2O_3/Fe^0 and the H_2O_2/Fe_3O_4/ Fe_2O_3/Fe^0 (500/500/ 3000, 1500/1500/ 1000, 500/500/1000, 125/125/750/ 375/ 375/250, 1000/1000/ 2000, 250/250/1500, 750/750/500, 250/ 250/500) pH = 2.0, 3.0, 4.0, 5.0, 6.0	H_2O_2/COD = 1:1 Time = 120 min Catalyst: 500/500/ 1000 mg/L Fe_3O_4/ Fe_2O_3/Fe^0 pH = 3.0	TOC = 134.1 mg/ L (56,2%)
[163]	Insights into the novel application of Fe-MOFs in ultrasound-assisted heterogeneous Fenton system	Fe-MOFs (MIL-53, MIL-88B and MIL-101)	Tetracycline hydrochloride	10 mg/L	XRD, FTIR, SEM, XPS, N_2 sorption-desorption isotherms and CO-FTIR, BET, TOC analyzer, ICP, UV-vis spectrometer, EPR	US power = 40, 60, 80, 100 W Catalyst = 0.2, 0.3, 0.4 g/L H_2O_2 = 22, 44, 88 mM pH = 5.0, 7.0, 9.0, 11.0	US Power = 60 W Catayst = 0.3 g/L H_2O_2 = 44 mM pH = 5.0–11.0 (efficiency not change)	TOC = 42.5% (after 7 minutes) for MIL88B

[164]	Decolorization of wastewater by heterogeneous Fenton reaction using MnO_2-Fe_3O_4/CuO hybrid catalysts	MnO_2-Fe_3O_4/CuO hybrid, Fe_3O_4/CuO	Methylene blue		XRD, EDS, FE-SEM, and BET techniques	H_2O_2 = 2, 4, 8, 16, 32 mmol/L pH = 2.0, 4.0, 5.0, 7.0, 9.5, 12.0 Catalyst = 10, 25, 50, 75, 100 mg/L	H_2O_2 = 16/32 mmol/L pH = 2.0/4.0 Catalyst = 10 mg/L	Methylene blue = 90% degradation for MnO_2-Fe_3O_4/CuO
[165]	Decolorization and mineralization of commercial reactive dyes by using homogeneous and heterogeneous Fenton and UV/Fenton processes	Fe^{2+} and Fe^0 concentrations, Fe^{2+}, Fe^0	Reactive Yellow 3 (RY3), C.I. Reactive Blue 2 (RB2), and C.I. Reactive Violet 2 (RV2)	100 mg/L	UV-vis Spectrophotometer, TOC analyzer	$FeSO_4{\cdot}7H_2O$ = 0.2, 0.5, 1.0, 2.0 mM H_2O_2 = 2.5, 5, 10, 20, 50 mM Fe^{2+}/H_2O_2 = 1/5, 1/10, 1/20, 1/40, 1/100 Fe^0/H_2O_2 = 1/0.2, 1/0.4, 1/2, 1/20, 1/200 H_2O_2 (on mineralization) = 0.01, 0.1, 1.0, 5.0, 10.0 mM Fe^0 = 0.1, 0.2, 0.4, 1.0, 2.0 mM Time = 0, 5, 10, 30, 40, 60, 70, 90 min	Fe^{2+}/H_2O_2 = 0.5 mM/20 mM Fe^0/H_2O_2 = 2 mM/1 mM	TOC = 78–84% Decolourization = 95–100%

DIVISION OF CATALYSTS IN THE FENTON PROCESS

In the heterogeneous Fenton process, iron catalysts are used to promote the oxidation of organic compounds in contaminated water and soil. Several different kinds of iron catalysts can be used in this process, including:

- Iron oxides: Iron oxides, such as goethite, hematite, and magnetite, are naturally occurring minerals that can act as catalysts in the Fenton reaction. These minerals are often found in iron-containing soils, and their high surface area and reactivity make them effective catalysts for the oxidation of organic pollutants.
- Iron-modified zeolites: Zeolites are porous materials that can be modified with iron ions to create an effective catalyst for the Fenton reaction. Iron-modified zeolites have a high surface area and reactivity and can be tailored to optimize their catalytic properties for specific applications.
- Iron-doped carbon materials: Carbon materials, such as activated carbon or carbon nanotubes, can be doped with iron ions to create a catalyst for the Fenton reaction. Iron-doped carbon materials have a high surface area and reactivity and can be easily synthesized and tailored to optimize their catalytic properties.
- Iron-based nanoparticles: Iron-based nanoparticles, such as zero-valent iron (ZVI) or iron oxide nanoparticles, can be used as catalysts in the Fenton reaction. These nanoparticles have a high surface area and reactivity and can be easily dispersed in water or soil for effective treatment of contaminated sites.
- Iron-based waste materials: Waste materials, such as blast furnace slag or steel slags, can be used as catalysts in the Fenton reaction. These materials contain high levels of iron, and their use as catalysts can provide an economical and environmentally friendly solution for the treatment of contaminated water and soil.

The choice of iron catalyst depends on several factors, including the type and concentration of contaminants present in the water or soil, the pH and temperature of the system, and the desired reaction rate and efficiency. Further research is needed to better understand the properties and limitations of different types of iron catalysts in the heterogeneous Fenton process and to develop methods for optimizing their use in environmental remediation [13].

ZERO-VALENT IRON

Zero-valent iron (ZVI) is a form of iron that has a high reactivity due to the lack of valence electrons in its outer shell. ZVI is often used as a catalyst in environmental remediation applications, including the treatment of contaminated water and soil.

When ZVI is used as a catalyst in the Fenton reaction, it can promote the oxidation of organic contaminants in water and soil by generating hydroxyl radicals. ZVI can also be used as a reducing agent to convert contaminants, such as heavy metals or chlorinated solvents, into less toxic or non-toxic forms through a process called reductive dechlorination. ZVI is often used in permeable reactive barriers (PRBs), which are barriers made of a reactive material, such as ZVI, placed in the subsurface to intercept and treat contaminated groundwater as it flows through. The PRB allows the contaminated water to come into contact with the ZVI, which can remove or transform the contaminants through a variety of chemical processes. ZVI has several advantages as a remediation technology, including its low cost, high reactivity, and compatibility with other remediation technologies. However, the use of ZVI also has some limitations, including the potential for passivation, which occurs when the ZVI becomes coated with corrosion products or other materials that limit its reactivity. The use of ZVI also requires careful consideration of site-specific factors, including the type and concentration of contaminants present, the hydrogeological conditions, and the potential for off-site impacts. Overall, ZVI is a promising and widely used remediation technology that can effectively treat contaminated water and soil through a variety of chemical processes. Further research is needed to better understand the properties and limitations of ZVI as a catalyst and to develop methods for optimizing its use in environmental remediation. The disadvantage of the catalyst is deactivation in neutral and alkaline conditions [14].

Zero valent iron is one of the most popular solid catalysts used in the heterogeneous Fenton process and is the zero-valent iron (ZVI), characterized by activity, environmental friendliness, cost-effective preparing [15], non-toxicity [16], as well as effectiveness to remove halogenated compounds [17,18], nitrate [19], phosphate [20], polycyclic aromatic hydrocarbons [21,22], heavy metals [19,23,24], arsenic [25], dyes [26,27], phenol [28–30], nitrobenzene, chlorophenol, pharmaceutical compounds, or wastewater from the food industry, mining, and during the remediation of land contaminated with oil derivatives [31].

Iron-Containing Soils

Iron-containing soils have been used as heterogeneous catalysts in Fenton-like processes for the treatment of contaminated water and soil. These soils have a high content of iron oxides, such as goethite, hematite, and magnetite, which can act as a source of Fe^{2+} and Fe^{3+} ions for the Fenton reaction.

In Fenton-like processes that use iron-containing soils as catalysts, the soil is typically first pretreated to remove any organic matter or other contaminants that may interfere with the catalytic activity. The soil is then activated by reducing the Fe^{3+} ions to Fe^{2+} ions using a reducing agent, such as sodium dithionite or sodium borohydride. The activated soil is then mixed with the contaminated water or soil, and hydrogen peroxide is added to initiate the Fenton reaction.

The use of iron-containing soils as catalysts in Fenton-like processes has several advantages. These soils are abundant and widely available, making them a low-cost and environmentally friendly alternative to synthetic catalysts. They also have a high surface area, which can increase the efficiency of the reaction. Additionally, the soil can act as a filter or adsorbent, removing some of the pollutants from the water or soil before they even reach the catalytic sites.

However, the use of iron-containing soils in Fenton-like processes also has some limitations. The catalytic activity of the soil can be affected by factors such as pH, temperature, and the presence of other ions or contaminants. The soil may also contain other minerals or compounds that can interfere with the Fenton reaction or form by-products. The stability and reusability of the soil as a catalyst may also be a concern, as the soil may become saturated with contaminants over time or lose its catalytic activity through repeated use.

Overall, iron-containing soils are an important and promising natural resource for the remediation of contaminated water and soil. Further research is needed to better understand the properties and limitations of these soils as catalysts for the Fenton reaction and to develop methods for optimizing their use in environmental remediation [32–34].

Iron Minerals and Iron (hydr)Oxides

In the heterogeneous Fenton process, compounds such as solid iron oxides can be easily prepared in the laboratory and used as catalysts. In addition, their nanoforms have strong adsorption and catalytic activities [7,35,36]. In the heterogeneous Fenton process, popular iron oxides used as a catalyst are magnetite [37], hematite [38], ferrihydrite [39], ferrites [40,41], goethite [38], and schorl [42].

Magnetite

Magnetite with the formula Fe_3O_4 consists of 73% iron, which makes it one of the highest iron-content minerals. It is composed of Fe^{2+} oxides and iron Fe^{3+} oxides, where there are two Fe_3O_4 ions per one Fe^{2+} ion. The properties of magnetite such as low price, no toxicity, and strong magnetic properties, guarantee its usage in environmental protection. It is also a popular catalyst in advanced oxidation processes (AOPs) due to its ability to remove pollutants based on redox reactions. In addition, we can separate magnetite from post-process sludge due to its magnetic properties. Magnetite owns its properties to many octahedral sites in its structure. As a result, it can hold many Fe^{2+} ions, which speeds up and increases the reaction for initiating H_2O_2 activation [37]. However, the usage of magnetite in pure form is limited due to a lower degree of oxidation when compared to iron soluble in the heterogeneous Fenton reaction [10]. Magnetite nanoparticles can be modified by an organic disulfide polymer (PTMT) and used in the adsorption of heavy metals coming from simulated high-salinity wastewater [43]. In other effective and innovative research, the photocatalytic ozonation process has been used. Magnetite and titania support onto graphene were applied for micropollutants removal [44]. Magnetite with zero-valent iron and hematite as a catalyst were used to remove pollutants in cosmetic wastewater. The best wastewater treatment results were obtained for 500 mg/L Fe_3O_4, 500 mg/L Fe_2O_3, and Fe^0 1,000 mg/L [45]. We also used magnetite with hematite as a catalyst in AOP applications for cosmetic sewage treatment plants. In this research, the heterogeneous Fenton process was efficient and innovative. What is more, we noticed high efficiency of wastewater treatment, such as 75.7% TOC removal and 90.5% total nitrogen removal [45].

Hematite

Hematite is one of the most durable forms of trivalent iron oxide with high thermodynamic stability. It is a common material of natural sediments and soil, which contains about 70% iron. Hematite is an inexpensive and crude material that is applied as a combustible material in pig iron production after the process. Due to burning it in a furnace, hematite does not adversely affect the environment. However, before using the material in research, hematite is subjected to crushing, dispersing, and cleaning. The specified particle size (less than 60 nm) is achieved using the sedimentation process. The iron sample is sonicated in 95% ethanol for 5 min, washed with 0.1 M HNO_3 and rinsed with water, and mixed with 95% ethanol. The last stage of hematite sample preparation is drying at 30°C [46]. The freshly precipitated and amorphous iron oxide such as ferrihydrite is a precursor of hematite [47]. Hematite has a lower ability to degrade hydrogen peroxide than other minerals with higher efficiency in catalytic decomposition and a high degree of diffusion. Among the used catalysts, hematite was distinguished by the highest activity in catalyzing the oxidation reaction. The iron with lower valence is oxidized the fastest by H_2O_2. Lei *et al.* [48] noticed that the oxidation of CO in nanopores is more effective due to the higher density of iron and exposed active surface involved in the reaction [38]. The exposed surface of the crystals, in which the reaction takes place, size, and morphology, the arrangement of atoms on the surface of the compound, and the number and type of hydroxyl groups bonding to iron atoms influence hematite catalytic activity. Nanostructures of compounds and their modification such as Nb-containing hematite [49], Nb-doped hematite [50], Fe_2O_3 core-shell nanowires [51], S-doped hematite [52], N-co-doped hematite [53], are increasingly used as catalysts in heterogeneous Fenton process.

Ferrihydrite

Ferrihydrite is an iron oxide mineral (oxy)hydrate most often used in the form of a nanomaterial with a large specific surface area (SSA > 200 m^2/g) and many Fe-OH groups. Mineral has a structural formula $5Fe_2O_3{\cdot}9H_2O$. It occurs naturally in the earth's crust in the form of hydrated Fe^{3+} hydroxide. Ferrihydrite has a large SSA of 250–275 m^2/gL as well as high reactivity [39], which allows greater contact between the ferrihydrite and hydrogen peroxide. It also has a nanometric structure, which enables its more excellent contact between reactants and impurities and is a reasonable basis for dispersing functional materials [54]. Using ferrihydrite as a catalyst in the heterogeneous Fenton process is more and more popular [39]. Matt *et al.* [55] confirmed the well activity of catalysts in the heterogeneous Fenton process. They noticed that ferrihydrite achieved higher removed of 2,4,6-trinitrotoluene than with used other compounds such as hematite as well as goethite as the catalyst. The ferrihydrite also can be modified, which was done by Yanping *et al.* [56]. They synthesized Ag/AgBr/Ferrihydrate due to placing various Ag/AgBr on the ferrihydrite surface and using it as a catalyst. Results of the research presented that in the process, shows that electrons accelerate the formation of Fe^{2+}. In the process, in which electrons increase the efficiency of using H_2O_2, scientists noticed extend of pH values. During Fe^{3+} reduction, electrons can increase the structural stability of catalysts [57,58]. Wu *et al.* [59] used another modification of the process with the additional UV radiation to obtain dissolved ferrihydrite. Barreiro *et al.* [60] used ferrihydrite as a catalyst to remove atrazine with H_2O_2. While Zhang *et al.* [61] used citrate-modified ferrihydrite microstructures as a catalyst, synthesized via a simple aqueous solution route. The modified ferrihydrite microstructures adsorption well methylene blue and Cr^{6+} ions. The authors noticed also high activity in the generation of hydroxyl radicals by the usage of these compounds as catalysts [61].

Ferrites

Another popular catalyst applied in the heterogeneous Fenton process is ferrite. It is a metal oxide of the general formula AB_2O_4 is spinel. The structural formula is as follows MFe_2O_4, where M = Mn, Fe, Co, Ni, Zn, etc. [7]. Magnetite is an iron ferrite, in which Fe is siding in A and B positions. Magnetite is used in industry as powder or ceramics in electronic devices, magnetic nanodevices, optics, biomedicine, and telecommunications [40,41]. The ferrite has a crystalline structure, magnetic properties, and favorable adsorption properties. The three spinel ferrite structures depend on the M^{2+} and Fe^{3+} site position in the MFe_2O_4 pattern such as normal structure with M^{2+} in tetrahedral places and Fe^{3+} in octagonal places. Another structure is an inverse pattern with evenly distributing Fe^{3+} and M^{2+} in octagonal places. In mixed structures of spinel ferrite, the ions occupy tetrahedral and octahedral sites randomly. Normal structure spinel ferrite is $ZnFe_2O_4$, an inverse spinel ferrite is $NiFe_2O_4$ and Fe_3O_4 and the mixed spin ferrite includes $MnFe_2O_4$ [62]. Scientists also utilized nickel ferrite [63], bismuth ferrite [64], mesoporous silica magnetically surrounded [65], transition metals substituted with cobalt nano ferrites [66], nickel doped cobalt ferrite [67], as the catalyst. Scientists also used ferrihydrite as a photocatalyst, due to noticed effect of UV on ferrihydrite. Jauhar *et al.* [68] noticed the high effectiveness of cobalt ferrite with the addition of manganese and used it as a catalyst in the removal of methyl blue even in the absence of light [68]. Based on the research of Wu *et. al.* [59], it was found that cobalt ferrites used as catalysts have a high treatment efficiency of anionic and cationic compounds. Liu *et al.* [63], Soltani *et al.* [64], Sahoo *et al.* [65], and Singh *et al.* [67] also used ferrite as catalysts. In all processes, an effective evaluation of degradation of the pollutant over 80% was achieved.

Goethite

Goethite, a common iron oxide mineral, has been found to be an effective catalyst for the Fenton process in wastewater treatment. Goethite is a naturally occurring mineral that is abundant and inexpensive. It has a high surface area and is capable of adsorbing a range of organic and inorganic pollutants from wastewater. Studies have shown that goethite can enhance the efficiency of the Fenton process in removing pollutants such as dyes, pesticides, phenols, and heavy metals. The goethite catalyst can be easily synthesized and has a longer lifespan compared to other iron-based catalysts, making it a promising candidate for industrial wastewater treatment. However, like other iron-based catalysts, goethite has limitations such as pH dependency, iron leaching, and the need for a high concentration of hydrogen peroxide. Researchers are still exploring ways to optimize the use of goethite in the

Fenton process to overcome these limitations and improve its efficiency and applicability in wastewater treatment.

Goethite is an iron oxyhydroxide with a surface area (of 100–200 m^2/g), thermodynamically stable at "room temperature". It rusts in almost all kinds of soils in the natural environment and other surface formations containing iron, which structure (α-FeOOH) consists of double strands edge dividing $FeO_3(OH)_3$ octahedrons. The construction of goethite consists of double fringes connected by dividing the corners and forming octagonal "tunnels" crossing hydrogen bridges. In the heterogeneous Fenton process, goethite is applied as a catalyst due to its parameters such as widespread occurrence, low price, and environmental friendliness. It occurs in water as a chemically active compound with low energy demand [69]. Goethite is slightly soluble, and its dissolution by the leaching process is caused by the proton in the Fenton reaction with Fe^{3+} (Eq. 3.1).

$$\alpha - FeOOH + 3H^{+} \rightarrow Fe^{3+} + 2H_2O \tag{3.1}$$

One of the advantages of using goethite as a catalyst in the Fenton process is that it can be used under a wide range of pH conditions. This is because goethite has a surface charge that allows it to be stable over a broad range of pH values, making it suitable for treating wastewater with varying pH levels. Another advantage is that goethite has a high selectivity towards the target pollutants, which means it can effectively remove the pollutants without causing further harm to the environment. This is because goethite has a specific surface area that allows it to selectively adsorb and oxidize the pollutants, leaving behind harmless by-products. Moreover, goethite can also be used in conjunction with other catalysts such as activated carbon, which can enhance the removal efficiency of pollutants in wastewater. This combination is effective in treating complex wastewater containing various contaminants. Overall, the use of goethite as a catalyst in the Fenton process has great potential for improving wastewater treatment processes. Its abundance, low cost, and high efficiency make it an attractive option for industrial wastewater treatment. However, further research is needed to optimize the use of goethite in the Fenton process and to investigate its long-term stability and environmental impact. After the removal of the organic compounds, the leaching process continues with Fe^{3+} by non-reducing dissolution with by-products. The dissolution of goethite increases during the degradation by-products of the process [70,71]. However, Lin *et al.* [72] showed no dissolving iron in basic conditions, which confirms the poor solubility of the compound.

Schorl

Schorl (chemical formula: $NaFe_3Al_6(BO_3)3\text{-}Si_6O_{18}(OH)_4)$ is a natural, complex borosilicate mineral containing iron, which is a tourmaline group, which confirms they are the trigonal space group R3m – static crystal structures [42]. The formula of tourmaline is XY_3Z_6 $[Si_6O_{18}]$ $[BO_3]$ W_4, where X = Ca, Na, K or vacancy; Y = Li, Mg, Fe^{2+}, Mn^{2+}, Al, Cr^{3+}, V^{3+}, Fe^{3+}; Z = Mg, Al, Fe^{3+}, V^{3+}, Cr^{3+}; and W = OH, F, O [73]. Spontaneous and permanent poles characterize the tourmaline. The electric dipoles, which guarantee an electric field on the surface of a compound, were noticed on the surface of poles [7,74], which is unique pyroelectric properties and piezoelectricity. Schorl is classified as terminal members, gravity (Y = Mg), schorl (Y = Fe^{2+}), tsilaisite (Y = Mn), olenite (Y = Al), and elbait (Y = Li, Al), depending on the occupancy of the Y and X sites [75]. The mineral can produce an electrostatic charge if subjected to a slight pressure or temperature change [42]. Schorl is a catalyst in many kinds of industries such as water treatment, air purification systems, and medicine. Schorl, a mineral of the tourmaline group, has also been investigated as a catalyst for Fenton-like reactions in wastewater treatment. Studies have shown that schorl has the potential to enhance the Fenton process by generating hydroxyl radicals that can effectively degrade organic pollutants. Schorl has a unique crystal structure that provides a large surface area for catalytic activity. It contains iron and other trace elements that can act as catalysts to generate hydroxyl radicals in the Fenton reaction. Schorl is effective in removing various pollutants, including dyes, phenols, and pharmaceuticals, from wastewater. The study reported that schorl had a higher catalytic activity than other iron-containing minerals such as magnetite and hematite in the Fenton process. Additionally, schorl is stable under acidic conditions, which is necessary for the Fenton reaction. However, further research is needed to optimize the use of schorl as a catalyst in the Fenton process for wastewater treatment. Other studies are required to understand the mechanism of catalytic activity and to determine the optimal conditions for schorl to enhance the Fenton process. Research has shown that schorl can effectively catalyze the Fenton reaction, leading to the degradation of a variety of pollutants in wastewater. In one study, schorl was used as a catalyst in the Fenton process to remove phenol from aqueous solutions. The results showed that schorl was effective in degrading phenol, achieving a degradation efficiency of 96% after just 15 minutes of reaction time. Schorl has also been investigated as a catalyst for the degradation of other pollutants in wastewater, such as dyes and pharmaceuticals. In one study, schorl was shown to be an effective catalyst for the degradation of the antibiotic sulfamethazine in the Fenton process. The study found that schorl achieved a degradation efficiency of 91% after 60 minutes of reaction time. Overall, schorl has shown promise as a catalyst in the Fenton process for the treatment of wastewater. Further research is needed to fully understand the potential of schorl and to optimize its use in the Fenton process [76–78].

Pyrite

Pyrite, also known as iron sulfide (FeS_2), is a mineral that is commonly found in natural environments. Pyrite has been studied in the context of the Fenton process due to its ability to catalyze the generation of hydroxyl radicals ($^{\bullet}OH$) in the presence of hydrogen peroxide (H_2O_2). The Fenton process with pyrite (FeS_2) involves the oxidation of Fe^{2+} to Fe^{3+} by $^{\bullet}OH$, which is generated by the reaction between H_2O_2 and pyrite. The oxidation of Fe^{2+} to Fe^{3+} produces Fe^{3+} ions, which can then react with H_2O_2 to generate more $^{\bullet}OH$. This process leads to a chain reaction that generates a large number of $^{\bullet}OH$, which can then react with pollutants to break them down into simpler, less harmful compounds. Pyrite can be used as a low-cost alternative to ferrous ions (Fe^{2+}) as a catalyst in the Fenton process. Pyrite has a high surface area and a high affinity for H_2O_2, which makes it a highly effective catalyst for the generation of $^{\bullet}OH$. In addition, pyrite is abundant and widely available, which makes it an attractive option for large-scale applications. However, the use of pyrite in the Fenton process can also present some challenges. Pyrite is a highly reactive mineral that can undergo rapid oxidation in the presence of air and water, which can lead to the formation of acidic conditions that can inhibit the Fenton process. In addition, the presence of other minerals in the natural environment can also affect the effectiveness of pyrite as a catalyst.

Overall, pyrite is an effective catalyst in the Fenton process, and its use can offer several advantages in terms of cost and availability. However, the use of pyrite in the Fenton process requires careful consideration of the environmental conditions and the presence of other minerals in the system. Pyrite has several advantages as a catalyst in the Fenton process. First, pyrite has a high surface area and a high affinity for H_2O_2, which makes it a highly effective catalyst for the generation of $^{\bullet}OH$. Second, pyrite is abundant and widely available in nature, which makes it a low-cost and sustainable option for large-scale applications of the Fenton process. Finally, pyrite is highly effective in the treatment of a wide range of pollutants, including organic compounds, dyes, and pharmaceuticals.

However, the use of pyrite in the Fenton process also presents some challenges. Pyrite is a highly reactive mineral that can undergo rapid oxidation in the presence of air and water, which can lead to the formation of acidic conditions that can inhibit the Fenton process. In addition, the presence of other minerals in the natural environment can also affect the effectiveness of pyrite as a catalyst.

To address these challenges, researchers have studied various methods for stabilizing pyrite and enhancing its effectiveness as a catalyst in the Fenton process. These methods include the use of organic coatings to protect pyrite from oxidation and the addition of surfactants to enhance the dispersion of pyrite in solution.

Overall, pyrite has shown great potential as a catalyst in the Fenton process, and its use can offer several advantages in terms of cost and sustainability. However, the use of pyrite in the Fenton process requires careful consideration of the environmental conditions and the presence of other minerals in the system to ensure its effectiveness as a catalyst. Ammar *et al.* used pyrite to decompose tyrosol in the electro-Fenton process and obtain about 90% mineralization of solutions [79–81].

Bimetallic

Various types of metallic catalysts characterized significant effectiveness in the Fenton process and could be used in industry [19]. A bimetallic material is one of modification ZVI, in which their structure settles on transition metals. Transition metals simplify the production of activated atomic hydrogen (H^+), strengthening the reducing agent characterized by accelerating iron corrosion. Also, bimetallic materials can create galvanic cells and increase the oxidation of ZVI [82]. Isomorphic iron is often replaced with other transition metals to increase the catalytic activity of the compound. One kind of modification uses oxygen vacancy (OV), *i.e.*, active places on the catalyst's surface, formed during adaptation to unequal substitutions or cation deficiency in the magnetite structure. Costa *et al.* [83] used oxygen vacancy and noticed the synergistic effect of a used compound on the decomposition of hydrogen peroxide during the usage of magnetite, Co, and Mn. Magalhães *et al.* [84] introduced chromium in the magnetite structure to produce for the first time a Cr–Fe spinel active heterogeneous Fenton system. It was noticed that chromium shows ionic width compatible with the magnetite spinel structure and high reactivity towards H_2O_2 activation [84]. Another factor is the importance of ions in preventing the recombination of photogenerated electrons (e^-) and holes h^+ on the catalyst surface [85]. Another kind of modification applied porous magnetic composites, characterized by a high degree of magnetism and catalytic property, containing electron-carrying composite structures and exhibiting large surfaces with active centers. Other advantages of mentioned modification are larger pores, the grid structure favoring the diffusion of pollutants on the activated carbon, and the improvement of active adsorption sites [86]. Ma *et al.* [87] used magnetic porous carbon bullets derived from microporous chelating resin with cheap and environmentally friendly material, modified with ethylenediaminetetraacetic acid (EDTA), and loaded with iron composite. Chelating agents are alternative compounds that increase the efficiency of the heterogeneous Fenton process [88–90]. Catalyst can also be modificative by doping magnetic metal on the surface of compounds, which improves the catalytic properties. Pouran *et al.* [91] modified the magnetite by deposition of niobium on the catalyst structure. The researchers noticed that with a higher niobium content and a molar content, the degradation degree of methylene blue has increased. Using niobium as a catalyst with greater adsorption capacity increases the degradation efficiency of hydroxyl peroxide, and causes smaller elation of Nb ions. Moreover, an increase in the amount of niobium caused the catalyst crystal sizes to be smaller, and the surface area was larger in the compound due to the deficiency of cations in the catalyst structure. After doubling the amount of introduced Nb and Nb^{5+} ions, scientists noticed that with the replacing of Fe^{3+} cations in the crystal structure, and increasing particle size. Comparing the surface area of Fe^{3+} and Nb^{5+}, niobium has a larger radius than iron (55 *vs.* 64 μm). Because of that, it is characterized by a larger active surface area than iron [92]. The treatment process can also be effective due to the participation of Nb^{4+}/Nb^{5+} redox pairs in the H_2O_2 decomposition cycle. This process increases because of the rate of Fe^{3+} reduction to Fe^{2+} [91].

Other advantages of the process include increasing the adsorption on the surface of the catalyst, enhancing the oxygen quantity on the catalyst surface, and the regeneration of Fe^{2+} cations. One can also use an interesting and innovative solution as iron modified with graphene as a catalyst (0–2 wt.%). The iron deposited begins in the form of FeOOH on a graphene sheet showing a wide pH value range, stability, and reusability. All catalysts have unique properties such as the mesoporous structure and the large room for adsorption. Also, graphene has excellent electron conductivity (4,000 K/Wm) and a positive effect on the transfer of electrons between molecules, which leads to a reduction of iron Fe^{3+} to Fe^{2+}. Qui *et al.* [93] and Yang *et al.* [94] confirmed that using of modificative catalyst, which caused the process, could be cheaper and easier to use [95].

The iron-carbon with polytetrafluoroethylene (PTFE) was used by Zhang *et al.* [96], who used Fe^{2+} as the source of iron generated by anodic oxidation of Fe-C during micro-electrolysis. Scientists used molecules, with low process prices and high catalytic properties, and the long life of a galvanic cell. The PTFE used to modify the surface properties of Fe-C (iron with carbon) caused the limitation of iron ions leaching. The modification achieved good process efficiency at an initially neutral pH value [97]. Another modification of the process conducted Luo *et al.* [98] used a two-electron reduction of dissolved oxygen with carbon nanotubes coated with Fe_2O_3 with nickel foam as the cathode substrate to degrade pollutants. The advantages of carbon usage were a rapid reduction of O_2 and accumulation of H_2O_2 due to the high mass transfer efficiency and large active surface [98].

Saratale *et al.* [99] used the aromatic hydroxamic acid derivative of 2-hydroxypyridine-N-oxide (HpO) to build a catalyst with iron-HpO ligand deposited on granular activated carbon (GAC). The catalyst consisted of acid groups and N-derivatives of hydroxamic products, including chelating products and Fe^{3+} bond insurance [100]. Additionally, scientists formed hydroxypyridine N-oxide by the stable bidentate complexes with Fe^{3+} [101]. Another modification carried out by Zhao *et al.* [102] used a bifunctional recite (FeR) modified with iron as a catalyst. Hussain *et al.* [103] used strontium-doped copper deposited on zirconium oxide ZrFeSr as a catalyst in the heterogeneous Fenton process. Zhou *et al.* [6] noticed decomposed pollutants in the soil treatment in the heterogeneous Fenton process due to the use of ZVI/EDTA/Air as catalysts. The hydrogen peroxide quickly decomposes during contact with soil. By the reaction of ZVI with air, iron precipitates under neutral or alkaline conditions. Therefore, scientists utilized chelated acids to increase the efficiency of the process. They also used EDTA in the reaction with Fe^{2+}. The effect of the process was the formation of H_2O_2 [6]. Zhou *et al.* [104] based their process on hydrogen peroxide production by activating dioxygen in the reaction of iron-EDTA ligands. In this modification, it is important to activate oxygen from the iron- EDTA ligand reaction to form the O_2-Fe^{2+}-EDTA complex, which has a coordination number of 7 with metal in the hub [104]. Also, excess iron can reduce EDTA degradation, which accelerates the decomposition of H_2O_2 and the higher use of oxidants. Moreover, too much EDTA can slow down the kinetic reactions, especially O_2 activation and H_2O_2 reduction. The excessive EDTA (molar ratio [EDTA]:[Fe^{2+}/Fe^{3+}] greater than 1:1) can reduce the coordination number and prevent the formation of O_2 ligands F-Fe^{2+}/Fe^{3+}-EDTA and/or O_2-Fe^{2+}/Fe^{3+}-EDTA. Aleksić *et al.* [105] used Fe in the form of ZSM_5 zeolite and noticed the higher efficiency of the process at a higher pH value. At a higher pH value, the leaching of iron on the zeolite support was smaller than in an acidic environment. As a result, they lost less catalyst, and therefore it can be used with better results. During the reaction of H_2O_2 with iron ions, it bound on the zeolite skeleton and diffuses with the resulting radicals penetrating the mass, mostly $^{\bullet}OH$ radicals being formed on the inner part of zeolites [106]. During the process at a higher (almost neutral) pH value, the reaction of positively charged iron with the ions of the negatively charged zeolite backbone slows down or delays the generation of hydro complexes with iron ions, which are stable in water [107]. Ma *et al.* [87] used NH_2OH to accelerate the redox reaction of Fe^{3+}/Fe^{2+}, which resulted in a relatively constant recovery of Fe^{2+}. The recovery of Fe^{2+} increased the pseudo-first-order reaction rate and the pH range in which the treatment process could be carried out [87].

Waste Materials

The Fenton-iron process is an effective method for the treatment of wastewater that contains organic pollutants. One of the key components of the Fenton-iron process is the use of iron catalysts, which can be obtained from various waste materials.

Iron-based waste materials such as iron filings, scrap iron, steel slag, and iron oxide nanoparticles have been used as catalysts in the Fenton-iron process. These waste materials are attractive for use as catalysts due to their low cost, abundance, and environmental benefits of waste recycling [108].

Iron filings are commonly used as catalysts in the Fenton-iron process due to their high reactivity and low cost. They can be obtained from various sources, including metalworking shops and waste recycling centers. Similarly, scrap iron and steel slag, which are by-products of the steel industry, can also be used as catalysts in the Fenton-iron process [108].

In addition, iron oxide nanoparticles have shown promise as effective catalysts for the Fenton-iron process. These nanoparticles can be synthesized from various waste materials, such as waste iron sludge or iron-rich wastewater, through simple and low-cost methods.

The use of waste materials as catalysts in the Fenton-iron process not only reduces the cost of the treatment process but also provides an environmentally friendly solution for waste recycling. It also helps to reduce the burden on landfills and promotes the circular economy concept [108].

However, it is important to note that the quality of the waste materials used as catalysts can affect the efficiency of the Fenton-iron process. The waste materials should be carefully selected and processed to ensure their purity and effectiveness as catalysts.

In conclusion, waste materials containing iron can be used as effective catalysts in the Fenton-iron process for the treatment of wastewater containing organic pollutants. This provides an environmentally friendly and cost-effective solution for waste recycling and the treatment of wastewater [109].

Waste iron materials such as iron filings, scrap iron, and steel slag can be used as catalysts in the Fenton-iron process for the treatment of wastewater. However, these materials may contain impurities that can affect their catalytic activity. Therefore, it is important to properly prepare the waste iron materials before use in the Fenton-iron process.

Here are some steps for preparing waste iron materials for use in the Fenton-iron process:

- Collect the waste iron materials: The first step is to collect the waste iron materials from various sources, such as metalworking shops and waste recycling centers.
- Clean the waste iron materials: The waste iron materials should be cleaned to remove any contaminants, such as oil, grease, or dirt. This can be done by washing the materials with water and soap and then rinsing them thoroughly with clean water.

- Dry the waste iron materials: The waste iron materials should be dried completely to prevent any residual moisture from affecting their catalytic activity. This can be done by air-drying the materials or by placing them in an oven at a low temperature.
- Grind the waste iron materials: The waste iron materials can be ground into smaller particles to increase their surface area and improve their catalytic activity. This can be done using a ball mill or a mortar and pestle.
- Test the catalytic activity: Finally, the prepared waste iron materials should be tested for their catalytic activity in the Fenton-iron process. This can be done by conducting batch experiments with different concentrations of hydrogen peroxide and iron catalysts and measuring the degradation of organic pollutants in the wastewater [108].

In conclusion, preparing waste iron materials for use in the Fenton-iron process requires careful cleaning, drying, grinding, and testing to ensure their effectiveness as catalysts. Proper preparation can help to improve the efficiency of the Fenton-iron process for the treatment of wastewater containing organic pollutants.

Waste iron materials can be obtained from various sources for use in the Fenton-iron process. Some common sources of waste iron include:

- Metalworking shops: Waste iron materials such as iron filings, scraps, and turnings are generated in metalworking shops as byproducts of metal processing and machining.
- Steel industry: The steel industry generates large amounts of waste iron materials, such as steel slag, which is a by-product of steel production.
- Waste recycling centers: Waste recycling centers collect and process various types of waste materials, including scrap iron and other metals, that can be used as catalysts in the Fenton-iron process.
- Construction sites: Construction sites generate waste iron materials such as steel reinforcement bars, which can be recycled and used in the Fenton-iron process.
- Landfills: Landfills contain a significant amount of waste iron materials that can be recycled and used in the Fenton-iron process. However, the quality of the waste materials obtained from landfills may be lower than those obtained from other sources [108].

In general, waste iron materials can be obtained from a wide range of sources, and their availability and quality may vary depending on the location and type of industry. It is important to properly select and prepare the waste iron materials for use in the Fenton-iron process to ensure their effectiveness as catalysts.

Overall, the use of waste iron materials as catalysts in the Fenton-iron process can be effective and efficient, with several advantages over other types of catalysts. Some of the advantages include:

- Low cost: Waste iron materials are readily available and inexpensive compared to other types of catalysts, such as noble metals.

- Environmental benefits: The use of waste iron materials in the Fenton-iron process promotes waste recycling and reduces the burden on landfills, thereby promoting a circular economy concept.
- High reactivity: Iron-based catalysts are highly reactive and can effectively degrade a wide range of organic pollutants in wastewater.
- Long-term stability: Iron-based catalysts are relatively stable and can maintain their catalytic activity over a long time, reducing the need for frequent replacement [108].

However, the efficiency of the Fenton-iron process using waste iron materials can be influenced by factors such as the particle size, purity, and concentration of the catalysts, as well as the operating conditions of the process, such as pH and temperature. Therefore, it is important to carefully select and prepare the waste iron materials for use in the Fenton-iron process to ensure their effectiveness as catalysts and optimize the treatment efficiency [108].

In the heterogeneous Fenton process, scientists utilized waste materials as a catalyst, thus saving money and natural materials. For example, Kumar *et al.* [110] used black soil from black iron-laden soil as a catalyst, which was prepared for the process by wet impregnation with iron (Fe-BS) and ticking at various temperatures in calcination. The calcination process was caused by the removal of volatile substances and impurities to increase the active surfaces on the catalyst and create large surfaces or active sites. Due to the various parameters of the process, especially the temperatures, the catalyst can be stiffer, more active, and more stable [111,112]. Mashayekh-Salehi *et al.* [113] in their research, used pyrite-induced waste rock from the Ahan-Lajaneh mine in Iran as a catalyst. They used nanoparticles with mesoporous powder, which are characterized by the presence of FeS_2 without additional compounds. Van *et al.* [114] used iron slag as a catalyst in the Fenton process from metallurgy processes in Thai Nguyen Non-Ferrous Metals Limited Company in Vietnam. Industrialists estimated producing of metal slag is about 250,000 tons per year. To prepare the catalyst, iron was dried at 105°C for 48 h. In the next stage, it was crushed and sieved. Therefore using waste in the Fenton process is cheaper and more friendly for the environment than their utilization. Chu *et al.* [115] used magnetic biochar derived from food as the catalyst. To prepare magnetic biochar scientists used food waste, for example from a food-waste disposal facility, which is located in Korea. The most abundant components of food were glucan (23.1%) and protein (21.5%). In food were also ash (16.6%) and other components (36.5%). They mixed Fe^{2+} with chloride tetrahydrate, Fe^{2+} with chloride hexahydrate, food waste, and distilled water to prepare the magnetic biochar. In the next step, pH was adjusted to 7.6–8.2 and dried in a furnace. The compound was pyrolyzed at 300°C for 7 h as well as dried in a vacuum oven [115]. Additionally, Hong-Chao *et al.* [116] used ionothermal carbonization of biomass to construct Fe, N-doped biochar with prominent activity and recyclability as cathodic catalysts in heterogeneous electro-Fenton (see Figure 3.1). Tan *et al.* [85] used waste iron oxide obtained from the tannery wastewater process as a catalyst. XRD analysis showed the goethite-compatible compound in the wastewater, which confirms the presence of iron-containing compounds [85]. Bansal *et al.* [117] used

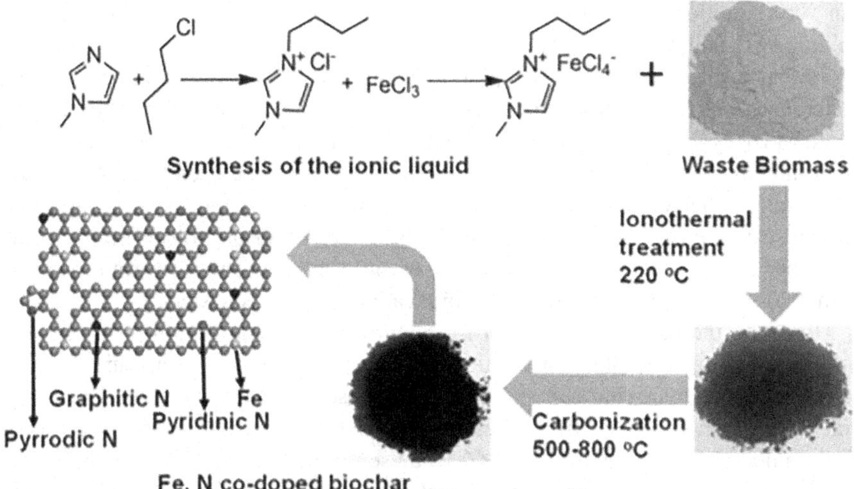

FIGURE 3.1 The schematic diagram for the synthesis of Fe, N cooped biochar materials.

Source: https://pubs.acs.org/doi/10.1021/acsestengg.0c00001.

remnants of car destruction such as the fly ash Fenton process with UV light as a catalyst for the removal of tetracycline. Usually, industrialists deposit about 5 million tons of automotive waste in landfills yearly. This type of waste has metals such as copper and aluminum in their structure, which are popularly used as catalysts in wastewater treatment [117].

Fly ash is a by-product of coal combustion in power plants and is a common waste material. Fly ash contains various elements and minerals, including iron, which can be used as a potential source of iron in the Fenton process. Fly ash iron has been investigated for its potential use as a catalyst in the Fenton process to degrade organic pollutants in wastewater.

Studies have shown that fly ash iron can be effective in generating hydroxyl radicals and degrading various types of organic pollutants. The use of fly ash iron in the Fenton process has several advantages. Firstly, fly ash is a low-cost and abundant waste material, making it an attractive option for wastewater treatment applications. Secondly, the use of fly ash iron can reduce the amount of waste material that needs to be disposed of, providing an environmentally sustainable solution. However, the use of fly ash iron in the Fenton process also has some limitations. The iron content of fly ash can vary significantly, which can affect its catalytic activity in the Fenton process. The presence of other elements and minerals in fly ash can also affect its catalytic activity and stability in the Fenton process.

To address these limitations, various methods have been proposed to prepare fly ash iron for use in the Fenton process. For example, acid leaching can be used to extract iron from fly ash and increase its iron content. Thermal treatment can also be used to enhance the catalytic activity of fly ash iron. In conclusion, fly ash iron is a potential source of iron in the Fenton process, and its use can provide a low-cost and sustainable solution for wastewater treatment. However, the iron content and properties of fly ash can vary significantly, and preparation methods should be carefully considered to optimize its catalytic activity and stability in the Fenton process.

Further research is needed to explore the potential of fly ash iron and other waste materials as catalysts in the Fenton process. In addition to its use as a catalyst in the Fenton process, fly ash iron has also been investigated for other applications in environmental remediation. For example, fly ash iron can be used to remove heavy metals and other contaminants from water and soil. The high surface area and porous structure of fly ash iron allow for the effective adsorption of heavy metals and other contaminants. The iron oxide present in fly ash can also form complexes with heavy metals, making them less bioavailable and less toxic.

However, the use of fly ash iron for environmental remediation also has some limitations. The variable composition and properties of fly ash can affect its effectiveness in removing contaminants. The presence of other elements and minerals in fly ash can also affect the stability and performance of fly ash iron in environmental remediation. To address these limitations, various modifications have been proposed to enhance the effectiveness of fly ash iron in environmental remediation. For example, fly ash can be modified through acid or alkaline treatments to increase its iron content and improve its adsorption capacity. Other modifications, such as the addition of organic matter, can also enhance the performance of fly ash iron in environmental remediation [112,118–120].

Iron sludge is a by-product generated from the iron-based coagulants used in wastewater treatment. Due to its high iron content, it has the potential to be used as a catalyst in the Fenton process for wastewater treatment. The iron sludge can act as a heterogeneous catalyst in the Fenton reaction, providing a high surface area for the reaction to occur and facilitating the production of hydroxyl radicals. Research has shown that the addition of iron sludge to wastewater can enhance the performance of the Fenton process. The iron sludge can act as a source of iron ions, which can participate in the Fenton reaction and promote the formation of hydroxyl radicals. Additionally, the use of iron sludge can reduce the amount of waste generated from the wastewater treatment process. However, the use of iron sludge in the Fenton process also presents some challenges. The quality and composition of the iron sludge can vary, which can affect its catalytic activity. In addition, the presence of other impurities in the iron sludge, such as heavy metals, can negatively impact the performance of the Fenton process and lead to environmental concerns. Overall, the use of iron sludge in the Fenton process shows promise as a way to enhance the efficiency and sustainability of wastewater treatment. Further research is needed to optimize the use of iron sludge as a catalyst and to address any environmental concerns associated with its use. Iron sludge, also known as ferric sludge or iron hydroxide, is a by-product generated in various wastewater treatment processes, such as coagulation/flocculation with ferric chloride or ferric sulfate. Iron sludge contains a high amount of iron, which makes it a potential candidate for use as an iron source in the Fenton process. Studies have shown that iron sludge can be used effectively as an iron source in the Fenton process for the treatment of various wastewater contaminants, such as dyes, phenols, and organic acids. However, the effectiveness of iron sludge as an iron source depends on several factors, including its iron content, the type of contaminants present in the wastewater, and the operating conditions of the Fenton process. One of the major advantages of using iron sludge in the Fenton process is that it is a low-cost source of iron, which can significantly reduce the overall cost of the Fenton process. Additionally, the use of iron sludge can also reduce the amount of waste generated from wastewater treatment processes, as it can be used as a value-added product instead of being disposed of as a waste material. However, the use of iron sludge in the Fenton process also has some limitations. For example, the composition of iron sludge can vary depending on the source and treatment process, which can affect its performance in the Fenton process. Moreover, the presence of impurities in iron sludge, such as heavy metals, can pose a risk to the environment and human health if not properly handled.

Overall, the use of iron sludge in the Fenton process has shown promise as an alternative to commercial iron salts for the treatment of wastewater contaminants. However, further research is needed to optimize its use and to evaluate its potential risks and benefits [13,121].

IRON CARRIERS IN CATALYSTS

Iron carriers used in the Fenton process can include various materials such as iron filings, scrap iron, iron oxide, iron hydroxide, iron carbonate, and iron-containing clays. These materials can act as carriers for iron particles and enhance the performance of the Fenton process. The choice of iron carrier depends on the specific wastewater treatment application and the properties required for the carrier. One of the commonly used iron carriers in the Fenton process is iron oxide, specifically magnetite and hematite. These materials have a high surface area, thermal stability, and resistance to acidic and basic environments, which makes them ideal carriers for the iron catalyst in the Fenton process. Iron oxide carriers can also provide a high concentration of iron particles, leading to enhanced catalytic activity in the degradation of organic pollutants [122,123].

Another common iron carrier in the Fenton process is iron-containing clays, such as montmorillonite and bentonite. These clays have a high surface area and cation exchange capacity, allowing them to effectively immobilize iron particles and enhance their reactivity in the Fenton process.

The use of iron carriers in the Fenton process can improve the stability and efficiency of the iron catalyst, leading to enhanced degradation of organic pollutants in wastewater. However, the choice of carrier material and its properties should be carefully considered to ensure optimal performance of the Fenton process.

In conclusion, iron carriers play an important role in the Fenton process by immobilizing or dispersing the iron catalyst and improving its stability and reactivity. Various iron carriers can be used in the Fenton process, including iron oxide and iron-containing clays, and their properties should be carefully considered to optimize the performance of the process [124,125].

In addition to the iron carriers mentioned above, other materials have been investigated for their potential use as iron carriers in the Fenton process. For example, activated carbon has been shown to effectively immobilize iron particles and enhance their reactivity in the Fenton process. Activated carbon has a high surface area and porous structure, which allows for effective adsorption and immobilization of iron particles.

Various modifications have also been made to iron carriers to enhance their performance in the Fenton process. For example, iron oxide nanoparticles have been coated with various materials, such as silica and chitosan, to improve their stability and reactivity in the Fenton process. Iron-containing materials have also been immobilized on various carrier materials, such as graphene oxide and carbon nanotubes, to improve their dispersibility and catalytic activity. It is worth noting that the choice of iron carrier and its properties can also affect the kinetics of the Fenton process. For example, the size and shape of iron particles can affect their reactivity and the rate of hydroxyl radical production. The concentration and distribution of iron particles on the carrier material can also affect their catalytic activity. Overall, the use of iron carriers in the Fenton process is an important strategy to enhance the performance and stability of the iron catalyst. The choice of carrier material and its properties should be carefully considered to optimize the performance of the Fenton process in the treatment of various types of wastewater. Further research is needed to explore the potential of new iron carriers and modifications to improve the efficiency of the Fenton process [124,125].

To reuse catalysts and increase the efficiency of the process scientists immobilized iron compounds on carriers or insoluble iron oxides such as goethite, magnetite, and hematite [126]. Microelectrolysis may take place during the heterogeneous Fenton process in which carbon is used as the catalyst carrier. Process microelectrolysis consists of mixing carbon or a composite of carbon or graphite with steel and iron. As a result of the process, microcells are formed, but the dominant process is adsorption [127–129]. The iron-based catalyst was also immobilized on a carrier such as silica [130], polymer [131], single silica batch [132], as well as clay [133] and clay oxides [134], or hydrogels [131,132,135].

Moreover, iron oxides were placed on commercial or developed materials such as FexOy - MWNT (multi-walled carbon nanotubes), Fe/C, Fe-ZSM_5 (Zeolite Socony Mobil-5), and Fe-Beta, to increase the active surface of the catalyst and facilitate its separation [136–138]. Another type of catalyst is mesoporous silicates, with long-range structural systems, such as SBA-15 molecular sieves, and a broad spectrum of activity. Using the modified compounds is popular in the absorption, catalysis, and separation process [139]. Benzaquén *et al.* [140] synthesized catalysts of mesoporous materials modified with iron (Fe-SBA-15 nanocomposites). Niveditha *et al.* [141] immobilized iron-on molybdophospahate ($FeMoPO_4$) with a tetrahedral structure to effectively form the catalyst. It was noticed that iron and silica are stable and effective catalysts. Moreover, silica has a lot of advantages such as the possibility of replacing it with phosphate [142]. Remirez *et al.* [136] noticed the adsorption process taking place in macropores. They analyzed it by the current microscope in the sample of coal from stone [136]. Another disadvantage is decreasing efficiency of the catalyst. During the recycling of Fe^{2+}/Fe^{3+}, sunlight is an essential element supporting the Fenton reaction. The use of a carrier can slightly limit light access to the catalyst during the regeneration of iron oxides [143].

REFERENCES

1. Oliveira, C., A. Alves, and L.M. Madeira, Treatment of water networks (waters and deposits) contaminated with chlorfenvinphos by oxidation with Fenton's reagent. *Chemical Engineering Journal*, 2014. **241**: p. 190–199.
2. Kong, S.-H., R.J. Watts, and J.-H. Choi, Treatment of petroleum-contaminated soils using iron mineral catalyzed hydrogen peroxide. *Chemosphere*, 1998. **37**(8): p. 1473–1482.
3. Wu, Q., et al., Copper/MCM-41 as catalyst for the wet oxidation of phenol. *Applied Catalysis B: Environmental*, 2001. **32**(3): p. 151–156.
4. Liou, R.-M., et al., Fe (III) supported on resin as effective catalyst for the heterogeneous oxidation of phenol in aqueous solution. *Chemosphere*, 2005. **59**(1): p. 117–125.
5. da Fonseca, F., et al., Heterogeneous Fenton process using the mineral hematite for the discolouration of a reactive dye solution. *Brazilian Journal of Chemical Engineering*, 2011. **28**: p. 605–616.
6. Zhou, H., et al., Removal of 2,4-dichlorophenol from contaminated soil by a heterogeneous ZVI/EDTA/Air Fenton-like system. *Separation and Purification Technology*, 2014. **132**: p. 346–353.
7. Nidheesh, P.V., Heterogeneous Fenton catalysts for the abatement of organic pollutants from aqueous solution: a review. *RSC Advances*, 2015. **5**(51): p. 40552–40577.
8. Morales-Torres, S., et al., Wet air oxidation of trinitrophenol with activated carbon catalysts: Effect of textural properties on the mechanism of degradation. *Applied Catalysis B: Environmental*, 2010. **100**(1): p. 310–317.
9. Apolinário, Â.C., et al., Wet air oxidation of nitro-aromatic compounds: Reactivity on single- and multi-component systems and surface chemistry studies with a carbon xerogel. *Applied Catalysis B: Environmental*, 2008. **84**(1): p. 75–86.
10. Neyens, E. and J. Baeyens, A review of classic Fenton's peroxidation as an advanced oxidation technique. *J Hazard Mater*, 2003. **98**(1-3): p. 33–50.
11. Muruganandham, M. and M. Swaminathan, Decolourisation of Reactive Orange 4 by Fenton and photo-Fenton oxidation technology. *Dyes and Pigments*, 2004. **63**(3): p. 315–321.
12. Bray, W.C. and M.H. Gorin, Ferryl Ion, A Compound of Tetravalent Iron. *Journal of the American Chemical Society*, 1932. **54**(5): p. 2124–2125.
13. Thomas, N., D.D. Dionysiou, and S.C. Pillai, Heterogeneous Fenton catalysts: A review of recent advances. *Journal of Hazardous Materials*, 2021. **404**: p. 124082.
14. Sun, Y.-P., et al., Characterization of zero-valent iron nanoparticles. *Advances in Colloid and Interface Science*, 2006. **120**(1): p. 47–56.
15. Michel, M., et al., Technological conditions for the coagulation of wastewater from cosmetic industry. *Journal of Ecological Engineering*, 2019. **20**: p. 78–85.
16. Yuan, Y., B. Lai, and Y.-Y. Tang, Combined Fe0/air and Fenton process for the treatment of dinitrodiazophenol (DDNP) industry wastewater. *Chemical Engineering Journal*, 2016. **283**: p. 1514–1521.
17. Wu, Y., et al., Characteristics and mechanisms of kaolinite-supported zero-valent iron/H2O2 system for nitrobenzene degradation. *CLEAN – Soil, Air, Water*, 2017. **45**(3): p. 1600826.
18. Zhang, H., et al., Pretreatment of shale gas drilling flowback fluid (SGDF) by the microscale Fe0/Persulfate/O3 process (mFe0/PS/O3). *Chemosphere*, 2017. **176**.
19. Ribeiro, A.R., et al., An overview on the advanced oxidation processes applied for the treatment of water pollutants defined in the recently launched Directive 2013/39/EU. *Environment International*, 2015. **75**: p. 33–51.
20. Wang, Z., et al., Removal of COD from landfill leachate by advanced Fenton process combined with electrolysis. *Separation and Purification Technology*, 2019. **208**: p. 3–11.

21. Yang, Z., et al., Enhanced Nitrobenzene reduction by zero valent iron pretreated with H2O2/HCl. *Chemosphere*, 2018. **197**: p. 494–501.
22. Barnes, R.J., et al., The impact of zero-valent iron nanoparticles on a river water bacterial community. *Journal of Hazardous Materials*, 2010. **184**(1): p. 73–80.
23. Li, X., et al., Characteristics and mechanisms of catalytic ozonation with Fe-shaving-based catalyst in industrial wastewater advanced treatment. *Journal of Cleaner Production*, 2019. **222**: p. 174–181.
24. Ebrahiem, E.E., M.N. Al-Maghrabi, and A.R. Mobarki, Removal of organic pollutants from industrial wastewater by applying photo-Fenton oxidation technology. *Arabian Journal of Chemistry*, 2017. **10**: p. S1674–S1679.
25. Bogacki, J., et al., Cosmetic wastewater treatment by the ZVI/H2O2 process. *Environmental Technology*, 2017. **38**(20): p. 2589–2600.
26. Zárate-Guzmán, A.I., et al., Towards understanding of heterogeneous Fenton reaction using carbon-Fe catalysts coupled to in-situ H2O2 electro-generation as clean technology for wastewater treatment. *Chemosphere*, 2019. **224**: p. 698–706.
27. Maojun, W., et al., The research on the relationship between industrial development and environmental pollutant emission. *Energy Procedia*, 2011. **5**: p. 555–561.
28. Zhang, Z., et al., The systematic adsorption of diclofenac onto waste red bricks functionalized with iron oxides. *Water*, 2018. **10**: p. 1343.
29. Xu, X., et al., Cyclohexanoic acid breakdown by two-step persulfate and heterogeneous Fenton-like oxidation. *Applied Catalysis B: Environmental*, 2018. **232**: p. 429–435.
30. Rezaei, F. and D. Vione, Effect of pH on zero valent iron performance in heterogeneous Fenton and Fenton-like processes: a review. *Molecules*, 2018. **23**(12): p. 3127.
31. Jan, B. and H. Al-Hazmi, Automotive fleet repair facility wastewater treatment using air/ZVI and air/ZVI/H2O2 processes. *Archives of Environmental Protection*, 2017. **43**.
32. Zhang, M.-h., et al., A review on Fenton process for organic wastewater treatment based on optimization perspective. *Science of The Total Environment*, 2019. **670**: p. 110–121.
33. Checa-Fernandez, A., et al., Application of chelating agents to enhance Fenton process in soil remediation: a review. *Catalysts*, 2021. **11**(6): p. 722.
34. Qin, J., et al., Decomposition of long-chain petroleum hydrocarbons by Fenton-like processes: Effects of ferrous iron source, salinity and temperature. *Ecotoxicology and Environmental Safety*, 2019. **169**: p. 764–769.
35. Han, Q., et al., Cobalt catalyzed peroxymonosulfate oxidation: a review of mechanisms and applications on degradating organic pollutants in water. *Progress in Chemistry*, 2012. **24**: p. 144–156.
36. Zhang, G., S. Wang, and F. Yang, Efficient adsorption and combined heterogeneous/homogeneous Fenton oxidation of amaranth using supported nano-FeOOH as cathodic catalysts. *The Journal of Physical Chemistry C*, 2012. **116**(5): p. 3623–3634.
37. Oturan, M.A., An ecologically effective water treatment technique using electrochemically generated hydroxyl radicals for in situ destruction of organic pollutants: Application to herbicide 2,4-D. *Journal of Applied Electrochemistry*, 2000. **30**(4): p. 475–482.
38. Chen, W., et al., Iron oxide containing graphene/carbon nanotube based carbon aerogel as an efficient E-Fenton cathode for the degradation of methyl blue. *Electrochimica Acta*, 2016. **200**: p. 75–83.
39. Wang, Y., et al., Photo-oxidation of Mordant Yellow 10 in aqueous dispersions of ferrihydrite and H2O2. *Journal of Molecular Catalysis A: Chemical*, 2010. **325**(1): p. 79–83.
40. He, J., et al., Interfacial mechanisms of heterogeneous Fenton reactions catalyzed by iron-based materials: A review. *Journal of Environmental Sciences*, 2016. **39**: p. 97–109.

41. Kharisov, B.I., H.V.R. Dias, and O.V. Kharissova, Mini-review: Ferrite nanoparticles in the catalysis. *Arabian Journal of Chemistry*, 2019. **12**(7): p. 1234–1246.
42. Barton, R., Refinement of the crystal structure of buergerite and the absolute orientation of tourmalines. *Acta Crystallographica Section B*, 1969. **25**.
43. Huang, X., et al., Design and synthesis of core–shell Fe3O4@PTMT composite magnetic microspheres for adsorption of heavy metals from high salinity wastewater. *Chemosphere*, 2018. **206**: p. 513–521.
44. Chávez, A.M., R.R. Solís, and F.J. Beltrán, Magnetic graphene TiO2-based photocatalyst for the removal of pollutants of emerging concern in water by simulated sunlight aided photocatalytic ozonation. *Applied Catalysis B: Environmental*, 2020. **262**: p. 118275.
45. Bogacki, J., et al., Magnetite, hematite and zero-valent iron as co-catalysts in advanced oxidation processes application for cosmetic wastewater treatment. *Catalysts*, 2021. **11**(1): p. 9.
46. Bel Hadjltaief, H., et al., Natural hematite and siderite as heterogeneous catalysts for an effective degradation of 4-chlorophenol via photo-Fenton process. *ChemEngineering*, 2018. **2**(3): p. 29.
47. Paterson, E., U. Schwertmann & R.M. Cornell. Iron Oxides in the Laboratory. Preparation and Characterization. VCH, Weinheim, 1991. xiv + 137 pp. Price £45 ISBN: 3.527.26991.6. *Clay Minerals*, 1992. **27**(3): p. 393–393.
48. Lei, L., et al., Oxidative degradation of poly vinyl alcohol by the photochemically enhanced Fenton reaction. *Journal of Photochemistry and Photobiology A: Chemistry*, 1998. **116**(2): p. 159–166.
49. Silva, A.C., et al., Nb-containing hematites Fe2–xNbxO3: The role of Nb5+ on the reactivity in presence of the H2O2 or ultraviolet light. *Applied Catalysis A: General*, 2009. **357**(1): p. 79–84.
50. Oliveira, H.S., et al., Nanostructured vanadium-doped iron oxide: catalytic oxidation of methylene blue dye. *New Journal of Chemistry*, 2015. **39**(4): p. 3051–3058.
51. Shi, J., Z. Ai, and L. Zhang, Fe@Fe2O3 core-shell nanowires enhanced Fenton oxidation by accelerating the Fe(III)/Fe(II) cycles. *Water Res*, 2014. **59**: p. 145–153.
52. Guo, L., et al., S-doped α-Fe2O3 as a highly active heterogeneous Fenton-like catalyst towards the degradation of acid orange 7 and phenol. *Applied Catalysis B: Environmental*, 2010. **96**(1-2): p. 162–168.
53. Pradhan, G.K., N. Sahu, and K.M. Parida, Fabrication of S, N co-doped α-Fe2O3 nanostructures: effect of doping, OH radical formation, surface area, [110] plane and particle size on the photocatalytic activity. *RSC Advances*, 2013. **3**(21): p. 7912–7920.
54. Xu, T., et al., Mechanisms for the enhanced photo-Fenton activity of ferrihydrite modified with BiVO4 at neutral pH. *Applied Catalysis B: Environmental*, 2017. **212**: p. 50–58.
55. Marcinowski, P., B. Jan, and J. Naumczyk, Oczyszczanie ścieków kosmetycznych z wykorzystaniem procesów koagulacji i Fentona The effectiveness of coagulation and Fenton process cosmetic wastewater treatment. *Gaz, woda; technika sanitarna*, 2014. **10**: p. 386–389.
56. Haber, F., J. Weiss, and W.J. Pope, The catalytic decomposition of hydrogen peroxide by iron salts. *Proceedings of the Royal Society of London. Series A - Mathematical and Physical Sciences*, 1934. **147**(861): p. 332–351.
57. Zhu, Y., et al., Visible-light Ag/AgBr/ferrihydrite catalyst with enhanced heterogeneous photo-Fenton reactivity via electron transfer from Ag/AgBr to ferrihydrite. *Applied Catalysis B: Environmental*, 2018. **239**: p. 280–289.
58. Xu, D., et al., Efficient removal of dye from an aqueous phase using activated carbon supported ferrihydrite as heterogeneous Fenton-like catalyst under assistance of

microwave irradiation. *Journal of the Taiwan Institute of Chemical Engineers*, 2016. **60**: p. 376–382.
59. Wu, Y., et al., Feasibility and mechanism of p-nitrophenol decomposition in aqueous dispersions of ferrihydrite and H2O2 under irradiation. *Reaction Kinetics, Mechanisms and Catalysis*, 2013. **110**.
60. Barreiro, J.C., et al., Oxidative decomposition of atrazine by a Fenton-like reaction in a H2O2/ferrihydrite system. *Water Res*, 2007. **41**(1): p. 55–62.
61. Zhang, X., et al., Citrate modified ferrihydrite microstructures: Facile synthesis, strong adsorption and excellent Fenton-like catalytic properties. *RSC Advances*, 2014. **4**: p. 21575.
62. Clarizia, L., et al., Homogeneous photo-Fenton processes at near neutral pH: A review. *Applied Catalysis B: Environmental*, 2017. **209**: p. 358–371.
63. Liu, S.-Q., et al., Magnetic nickel ferrite as a heterogeneous photo-Fenton catalyst for the degradation of rhodamine B in the presence of oxalic acid. *Chemical Engineering Journal*, 2012. **203**: p. 432–439.
64. Soltani, T. and M.H. Entezari, Solar-Fenton catalytic degradation of phenolic compounds by impure bismuth ferrite nanoparticles synthesized via ultrasound. *Chemical Engineering Journal*, 2014. **251**: p. 207–216.
65. Sahoo, B., et al., Fabrication of magnetic mesoporous manganese ferrite nanocomposites as efficient catalyst for degradation of dye pollutants. *Catalysis Science & Technology*, 2012. **2**(7): p. 1367–1374.
66. Jauhar, S. and S. Singhal, Substituted cobalt nano-ferrites, CoMxFe2−xO4 (M=Cr3+, Ni2+, Cu2+, Zn2+; 0.2≤x≤1.0) as heterogeneous catalysts for modified Fenton's reaction. *Ceramics International*, 2014. **40**: p. 11845–11855.
67. Singh, C., A. Goyal, and S. Singhal, Nickel-doped cobalt ferrite nanoparticles: efficient catalysts for the reduction of nitroaromatic compounds and photo-oxidative degradation of toxic dyes. *Nanoscale*, 2014. **6**(14): p. 7959–7970.
68. Jauhar, S., S. Singhal, and M. Dhiman, Manganese substituted cobalt ferrites as efficient catalysts for H2O2 assisted degradation of cationic and anionic dyes: Their synthesis and characterization. *Applied Catalysis A: General*, 2014. **486**: p. 210–218.
69. Wu, H., et al., Decolourization of the azo dye Orange G in aqueous solution via a heterogeneous Fenton-like reaction catalysed by goethite. *Environmental technology*, 2011. **33**: p. 1545–1552.
70. Lu, M.C., J.N. Chen, and H.H. Huang, Role of goethite dissolution in the oxidation of 2-chlorophenol with hydrogen peroxide. *Chemosphere*, 2002. **46**(1): p. 131–136.
71. Guimarães, I.R., et al., Modified goethites as catalyst for oxidation of quinoline: Evidence of heterogeneous Fenton process. *Applied Catalysis A: General*, 2008. **347**(1): p. 89–93.
72. Lin, S.-S. and M.D. Gurol, Catalytic decomposition of hydrogen peroxide on iron oxide: kinetics, mechanism, and implications. *Environmental Science & Technology*, 1998. **32**(10): p. 1417–1423.
73. Yavuz, F., A.H. Gültekin, and M.Ç. Karakaya, CLASTOUR: a computer program for classification of the minerals of the tourmaline group. *Computers & Geosciences*, 2002. **28**(9): p. 1017–1036.
74. Zong-Zhe, J., et al., Observation of spontaneous polarization of tourmaline. *Chinese Physics*, 2003. **12**(2): p. 222–225.
75. Prasad, P.S.R., Study of structural disorder in natural tourmalines by infrared spectroscopy. *Gondwana Research*, 2005. **8**(2): p. 265–270.
76. Proceedings First International Symposium on Environmentally Conscious Design and Inverse Manufacturing. in Proceedings First International Symposium on Environmentally Conscious Design and Inverse Manufacturing. 1999.

77. Xu, H.-y., M. Prasad, and Y. Liu, Schorl: A novel catalyst in mineral-catalyzed Fenton-like system for dyeing wastewater discoloration. *Journal of Hazardous Materials*, 2009. **165**(1): p. 1186–1192.
78. Xu, H., et al., Role of schorl's electrostatic field in discoloration of methyl orange wastewater using schorl as catalyst in the presence of H2O2. *Science China Technological Sciences*, 2010. **53**: p. 3014–3019.
79. Ammar, S., et al., Degradation of tyrosol by a novel electro-Fenton process using pyrite as heterogeneous source of iron catalyst. *Water Research*, 2015. **74**: p. 77–87.
80. Fathinia, S., et al., Preparation of natural pyrite nanoparticles by high energy planetary ball milling as a nanocatalyst for heterogeneous Fenton process. *Applied Surface Science*, 2015. **327**: p. 190–200.
81. Craig, J.R., F.M. Vokes, and T.N. Solberg, Pyrite: physical and chemical textures. *Mineralium Deposita*, 1998. **34**(1): p. 82–101.
82. Yamaguchi, R., et al., Hydroxyl radical generation by zero-valent iron/Cu (ZVI/Cu) bimetallic catalyst in wastewater treatment: Heterogeneous Fenton/Fenton-like reactions by Fenton reagents formed in-situ under oxic conditions. *Chemical Engineering Journal*, 2018. **334**: p. 1537–1549.
83. Costa, R.C.C., et al., Remarkable effect of Co and Mn on the activity of Fe3−xMxO4 promoted oxidation of organic contaminants in aqueous medium with H2O2. *Catalysis Communications*, 2003. **4**(10): p. 525–529.
84. Magalhães, F., et al., Cr-containing magnetites Fe3−xCrxO4: The role of Cr3+ and Fe2+ on the stability and reactivity towards H2O2 reactions. *Applied Catalysis A: General*, 2007. **332**(1): p. 115–123.
85. Tan, W., et al., Enhanced mineralization of Reactive Black 5 by waste iron oxide via photo-Fenton process. *Research on Chemical Intermediates*, 2020. **46**.
86. Zhou, L., et al., Fabrication of magnetic carbon composites from peanut shells and its application as a heterogeneous Fenton catalyst in removal of methylene blue. *Applied Surface Science*, 2015. **324**: p. 490–498.
87. Ma, J., et al., Novel magnetic porous carbon spheres derived from chelating resin as a heterogeneous Fenton catalyst for the removal of methylene blue from aqueous solution. *Journal of Colloid and Interface Science*, 2015. **446**: p. 298–306.
88. Bennedsen, L.R., et al., Mobilization of metals during treatment of contaminated soils by modified Fenton's reagent using different chelating agents. *Journal of Hazardous Materials*, 2012. **199-200**: p. 128–134.
89. Vicente, F., et al., Improvement soil remediation by using stabilizers and chelating agents in a Fenton-like process. *Chemical Engineering Journal*, 2011. **172**(2): p. 689–697.
90. Kang, N. and I. Hua, Enhanced chemical oxidation of aromatic hydrocarbons in soil systems. *Chemosphere*, 2005. **61**(7): p. 909–922.
91. Rahim Pouran, S., et al., Niobium substituted magnetite as a strong heterogeneous Fenton catalyst for wastewater treatment. *Applied Surface Science*, 2015. **351**: p. 175–187.
92. Robbins, M., et al., Magnetic properties and site distributions in the system FeCr2O4-Fe3O4, (Fe2+ Cr2-xFe3+xO4). *Journal de Physique Colloques*, 1971. **32**(C1): p. C1-266–C1-267.
93. Qiu, B., M. Xing, and J. Zhang, Mesoporous TiO2 nanocrystals grown in situ on graphene aerogels for high photocatalysis and lithium-ion batteries. *Journal of the American Chemical Society*, 2014. **136**(16): p. 5852–5855.
94. Yang, X., et al., Rapid degradation of methylene blue in a novel heterogeneous Fe3O4 @rGO@TiO2-catalyzed photo-Fenton system. *Sci Rep*, 2015. **5**: p. 10632.

95. Guo, S., et al., Graphene modified iron sludge derived from homogeneous Fenton process as an efficient heterogeneous Fenton catalyst for degradation of organic pollutants. *Microporous and Mesoporous Materials*, 2017. **238**: p. 62–68.
96. Zhang, C., et al., Heterogeneous electro-Fenton using modified iron–carbon as catalyst for 2,4-dichlorophenol degradation: Influence factors, mechanism and degradation pathway. *Water Research*, 2015. **70**: p. 414–424.
97. Zhang, C., et al., A new type of continuous-flow heterogeneous electro-Fenton reactor for Tartrazine degradation. *Separation and Purification Technology*, 2019. **208**: p. 76–82.
98. Luo, T., et al., Efficient degradation of tetracycline by heterogeneous electro-Fenton process using Cu-doped Fe@Fe2O3: Mechanism and degradation pathway. *Chemical Engineering Journal*, 2020. **382**: p. 122970.
99. Saratale, R.G., et al., Hydroxamic acid mediated heterogeneous Fenton-like catalysts for the efficient removal of Acid Red 88, textile wastewater and their phytotoxicity studies. *Ecotoxicol Environ Saf*, 2019. **167**: p. 385–395.
100. Kakkar, R., *Theoretical studies on hydroxamic acids*, in Hydroxamic Acids: A Unique Family of Chemicals with Multiple Biological Activities, S.P. Gupta, Editor. 2013, Springer Berlin Heidelberg: Berlin, Heidelberg. p. 19–53.
101. Sun, L., et al., Efficient removal of dyes using activated carbon fibers coupled with 8-hydroxyquinoline ferric as a reusable Fenton-like catalyst. *Chemical Engineering Journal*, 2014. **240**: p. 413–419.
102. Zhao, X., et al., Removing organic contaminants with bifunctional iron modified rectorite as efficient adsorbent and visible light photo-Fenton catalyst. *J Hazard Mater*, 2012. **215–216**: p. 57–64.
103. Hussain, S., et al., Enhanced ibuprofen removal by heterogeneous-Fenton process over Cu/ZrO2 and Fe/ZrO2 catalysts. *Journal of Environmental Chemical Engineering*, 2020. **8**(1): p. 103586.
104. Zhou, T., et al., Rapid decolorization and mineralization of simulated textile wastewater in a heterogeneous Fenton like system with/without external energy. *Journal of Hazardous Materials*, 2009. **165**(1): p. 193–199.
105. Aleksić, M., et al., Heterogeneous Fenton type processes for the degradation of organic dye pollutant in water – the application of zeolite assisted AOPs. *Desalination*, 2010. **257**: p. 22–29.
106. Neamţu, M., C. Catrinescu, and A. Kettrup, Effect of dealumination of iron (III)—exchanged Y zeolites on oxidation of Reactive Yellow 84 azo dye in the presence of hydrogen peroxide. *Applied Catalysis B: Environmental*, 2004. **51**(3): p. 149–157.
107. Feng, J., X. Hu, and P.L. Yue, Effect of initial solution pH on the degradation of Orange II using clay-based Fe nanocomposites as heterogeneous photo-Fenton catalyst. *Water Research*, 2006. **40**(4): p. 641–646.
108. Ali, M.E.M., T.A. Gad-Allah, and M.I. Badawy, Heterogeneous Fenton process using steel industry wastes for methyl orange degradation. *Applied Water Science*, 2013. **3**(1): p. 263–270.
109. Sathe, S.M., et al., Waste-derived iron catalyzed bio-electro-Fenton process for the cathodic degradation of surfactants. *Environmental Research*, 2022. **212**: p. 113141.
110. Kumar, V., et al., Degradation of mixed dye via heterogeneous Fenton process: Studies of calcination, toxicity evaluation, and kinetics. *Water Environ Res*, 2020. **92**(2): p. 211–221.
111. Gayapan, K., et al., Effects of calcination and pretreatment temperatures on the catalytic activity and stability of H 2 -treated WO 3 /SiO 2 catalysts in metathesis of ethylene and 2-butene. *RSC Advances*, 2018. **8**: p. 28555–28568.

112. Wang, N., et al., Study on preparation conditions of coal fly ash catalyst and catalytic mechanism in a heterogeneous Fenton-like process. *RSC Advances*, 2017. **7**: p. 52524–52532.
113. Mashayekh-Salehi, A., et al., Use of mine waste for H2O2-assisted heterogeneous Fenton-like degradation of tetracycline by natural pyrite nanoparticles: Catalyst characterization, degradation mechanism, operational parameters and cytotoxicity assessment. *Journal of Cleaner Production*, 2021. **291**: p. 125235.
114. Van, H.T., et al., Heterogeneous Fenton oxidation of paracetamol in aqueous solution using iron slag as a catalyst: Degradation mechanisms and kinetics. *Environmental Technology & Innovation*, 2020. **18**: p. 100670.
115. Chu, J.-H., et al., Application of magnetic biochar derived from food waste in heterogeneous sono-Fenton-like process for removal of organic dyes from aqueous solution. *Journal of Water Process Engineering*, 2020. **37**: p. 101455.
116. Li, H.-C., et al., Ionothermal carbonization of biomass to construct Fe, N-doped biochar with prominent activity and recyclability as cathodic catalysts in heterogeneous electro-Fenton. *ACS ES&T Engineering*, 2021. **1**(1): p. 21–31.
117. Bansal, P., T.S. Bui, and B.-K. Lee, Potential applications of ASR fly ash in photo-Fenton like process for the degradation of tetracycline at neutral pH: Fixed-bed approach. *Chemical Engineering Journal*, 2020. **391**: p. 123509.
118. Ribeiro, J.P., et al., Granulated biomass fly ash coupled with fenton process for pulp and paper wastewater treatment. *Environmental Pollution*, 2023. **317**: p. 120777.
119. Wang, N., et al., Treatment of polymer-flooding wastewater by a modified coal fly ash-catalysed Fenton-like process with microwave pre-enhancement: System parameters, kinetics, and proposed mechanism. *Chemical Engineering Journal*, 2021. **406**: p. 126734.
120. Wang, N., Q. Zhao, and A. Zhang, Catalytic oxidation of organic pollutants in wastewater via a Fenton-like process under the catalysis of HNO3-modified coal fly ash. *RSC Advances*, 2017. **7**(44): p. 27619–27628.
121. Bolobajev, J., et al., Reuse of ferric sludge as an iron source for the Fenton-based process in wastewater treatment. *Chemical Engineering Journal*, 2014. **255**: p. 8–13.
122. Gou, Y., et al., Degradation of fluoroquinolones in homogeneous and heterogeneous photo-Fenton processes: A review. *Chemosphere*, 2021. **270**: p. 129481.
123. Demarchis, L., et al., Photo–Fenton reaction in the presence of morphologically controlled hematite as iron source. *Journal of Photochemistry and Photobiology A: Chemistry*, 2015. **307-308**: p. 99–107.
124. Ganiyu, S.O., M. Zhou, and C.A. Martínez-Huitle, Heterogeneous electro-Fenton and photoelectro-Fenton processes: A critical review of fundamental principles and application for water/wastewater treatment. *Applied Catalysis B: Environmental*, 2018. **235**: p. 103–129.
125. Das, A. and M.K. Adak, Photo-catalyst for wastewater treatment: A review of modified Fenton, and their reaction kinetics. *Applied Surface Science Advances*, 2022. **11**: p. 100282.
126. Herrera, F., et al., Catalytic decomposition of the reactive dye UNIBLUE a on hematite. modeling of the reactive surface. *Water Research*, 2001. **35**(3): p. 750–760.
127. Amara, D., J. Grinblat, and S. Margel, Synthesis of magnetic iron and iron oxide micrometre-sized composite particles of narrow size distribution by annealing iron salts entrapped within uniform porous poly(divinylbenzene) microspheres. *Journal of Materials Chemistry*, 2010. **20**(10): p. 1899–1906.
128. Amara, D. and S. Margel, Synthesis and characterization of superparamagnetic core–shell micrometre-sized particles of narrow size distribution by a swelling process. *Journal of Materials Chemistry*, 2012. **22**(18): p. 9268–9276.

129. Liu, Y. and H. Yan, Carbon microspheres with embedded magnetic iron oxide nanoparticles. *Materials Letters*, 2011. **65**(7): p. 1063–1065.
130. Degradation of azo dye Reactive Black B using an immobilized iron oxide in a batch photo-fluidized bed reactor. *Environmental Engineering Science*, 2010. **27**(12): p. 1043–1048.
131. González-Bahamón, L.F., et al., Photo-Fenton degradation of resorcinol mediated by catalysts based on iron species supported on polymers. *Journal of Photochemistry and Photobiology A: Chemistry*, 2011. **217**(1): p. 201–206.
132. Cornu, C., et al., Identification and location of Iron species in Fe/SBA-15 catalysts: interest for catalytic Fenton reactions. *The Journal of Physical Chemistry C*, 2012. **116**(5): p. 3437–3448.
133. Hsueh, C.L., Y.H. Huang, and C.Y. Chen, Novel activated alumina-supported iron oxide-composite as a heterogeneous catalyst for photooxidative degradation of reactive black 5. *Journal of Hazardous Materials*, 2006. **129**(1): p. 228–233.
134. Rodríguez-Gil, J.L., et al., Heterogeneous photo-Fenton treatment for the reduction of pharmaceutical contamination in Madrid rivers and ecotoxicological evaluation by a miniaturized fern spores bioassay. *Chemosphere*, 2010. **80**(4): p. 381–388.
135. Huang, W., et al., Effect of ethylenediamine-N,N'-disuccinic acid on Fenton and photo-Fenton processes using goethite as an iron source: optimization of parameters for bisphenol A degradation. *Environ Sci Pollut Res Int*, 2013. **20**(1): p. 39–50.
136. Ramirez, J.H., et al., Azo-dye Orange II degradation by heterogeneous Fenton-like reaction using carbon-Fe catalysts. *Applied Catalysis B: Environmental*, 2007. **75**(3): p. 312–323.
137. Variava, M.F., T.L. Church, and A.T. Harris, Magnetically recoverable FexOy–MWNT Fenton's catalysts that show enhanced activity at neutral pH. *Applied Catalysis B: Environmental*, 2012. **123-124**: p. 200–207.
138. Gonzalez-Olmos, R., et al., Indications of the reactive species in a heterogeneous Fenton-like reaction using Fe-containing zeolites. *Applied Catalysis A: General*, 2011. **398**(1): p. 44–53.
139. Zhao, D., et al., Triblock copolymer syntheses of mesoporous silica with periodic 50 to 300 angstrom pores. *Science*, 1998. **279**(5350): p. 548–552.
140. Benzaquén, T.B., et al., Heterogeneous Fenton reaction for the treatment of ACE in residual waters of pharmacological origin using Fe-SBA-15 nanocomposites. *Molecular Catalysis*, 2020. **481**: p. 110239.
141. Niveditha, S. and R. Gandhimathi, Mineralization of stabilized landfill leachate by heterogeneous Fenton process with RSM optimization. *Separation Science and Technology*, 2020: p. 1–10.
142. Zhong, X., et al., Mesoporous silica iron-doped as stable and efficient heterogeneous catalyst for the degradation of C.I. Acid Orange 7 using sono–photo-Fenton process. *Separation and Purification Technology*, 2011. **80**(1): p. 163–171.
143. Wang, C., H. Liu, and Z. Sun, Heterogeneous photo-Fenton reaction catalyzed by nanosized iron oxides for water treatment. *International Journal of Photoenergy*, 2012. **2012**: p. 801694.
144. Barndõk, H., et al., Heterogeneous photo-Fenton processes using zero valent iron microspheres for the treatment of wastewaters contaminated with 1,4-dioxane. *Chemical Engineering Journal*, 2016. **284**: p. 112–121.
145. Dantas, T.L.P., et al., Treatment of textile wastewater by heterogeneous Fenton process using a new composite Fe2O3/carbon. *Chemical Engineering Journal*, 2006. **118**(1): p. 77–82.
146. Chakinala, A.G., et al., Industrial wastewater treatment using hydrodynamic cavitation and heterogeneous advanced Fenton processing. *Chemical Engineering Journal*, 2009. **152**(2): p. 498–502.

147. Hou, B., et al., Three-dimensional heterogeneous electro-Fenton oxidation of biologically pretreated coal gasification wastewater using sludge derived carbon as catalytic particle electrodes and catalyst. *Journal of the Taiwan Institute of Chemical Engineers*, 2016. **60**: p. 352–360.
148. Tu, Y., et al., Heterogeneous photo-Fenton oxidation of Acid Orange II over iron–sewage sludge derived carbon under visible irradiation. *Journal of Chemical Technology & Biotechnology*, 2014. **89**(4): p. 544–551.
149. Karthikeyan, S., et al., Treatment of textile wastewater by homogeneous and heterogeneous Fenton oxidation processes. *Desalination*, 2011. **281**: p. 438–445.
150. Sreeja, P.H. and K.J. Sosamony, A comparative study of homogeneous and heterogeneous photo-Fenton process for textile wastewater treatment. *Procedia Technology*, 2016. **24**: p. 217–223.
151. Liu, F., et al., Application of heterogeneous photo-Fenton process for the mineralization of imidacloprid containing wastewater. *Environmental Technology*, 2020. **41**(5): p. 539–546.
152. Ban, F.C., X.T. Zheng, and H.Y. Zhang, Photo-assisted heterogeneous Fenton-like process for treatment of PNP wastewater. *Journal of Water, Sanitation and Hygiene for Development*, 2020. **10**(1): p. 136–145.
153. Punzi, M., B. Mattiasson, and M. Jonstrup, Treatment of synthetic textile wastewater by homogeneous and heterogeneous photo-Fenton oxidation. *Journal of Photochemistry and Photobiology A: Chemistry*, 2012. **248**: p. 30–35.
154. Sun, L., Y. Li, and A. Li, Treatment of actual chemical wastewater by a heterogeneous Fenton process using natural pyrite. *Int J Environ Res Public Health*, 2015. **12**(11): p. 13762–13778.
155. Hu, J., et al., In-situ Fe-doped g-C3N4 heterogeneous catalyst via photocatalysis-Fenton reaction with enriched photocatalytic performance for removal of complex wastewater. *Applied Catalysis B: Environmental*, 2019. **245**: p. 130–142.
156. Bel Hadjltaief, H., et al., Influence of Operational Parameters in the Heterogeneous Photo-Fenton Discoloration of Wastewaters in the Presence of an Iron-Pillared Clay. Industrial & Engineering Chemistry Research, 2013. **52**(47): p. 16656–16665.
157. Yang, S., et al., Degradation of methylene blue by heterogeneous Fenton reaction using titanomagnetite at neutral pH values: process and affecting factors. *Industrial & Engineering Chemistry Research*, 2009. **48**(22): p. 9915–9921.
158. Jain, B., et al., Treatment of pharmaceutical wastewater by heterogeneous Fenton process: an innovative approach. *Nanotechnology for Environmental Engineering*, 2020. **5**(2): p. 13.
159. Xin, S., et al., High efficiency heterogeneous Fenton-like catalyst biochar modified CuFeO2 for the degradation of tetracycline: Economical synthesis, catalytic performance and mechanism. *Applied Catalysis B: Environmental*, 2021. **280**: p. 119386.
160. Lv, Y., et al., Remediation of organic arsenic contaminants with heterogeneous Fenton process mediated by SiO(2)-coated nano zero-valent iron. *Environ Sci Pollut Res Int*, 2020. **27**(11): p. 12017–12029.
161. Heidari, Z., et al., Application of mineral iron-based natural catalysts in electro-Fenton process: a comparative study. *Catalysts*, 2021. **11**(1): p. 57.
162. Parolini, M., A. Pedriali, and A. Binelli, Application of a biomarker response index for ranking the toxicity of five pharmaceutical and personal care products (PPCPs) to the bivalve Dreissena polymorpha. *Arch Environ Contam Toxicol*, 2013. **64**(3): p. 439–447.

163. Geng, N., et al., Insights into the novel application of Fe-MOFs in ultrasound-assisted heterogeneous Fenton system: Efficiency, kinetics and mechanism. *Ultrasonics Sonochemistry*, 2021. **72**: p. 105411.
164. Ghasemi, H., et al., Decolorization of wastewater by heterogeneous Fenton reaction using MnO2-Fe3O4/CuO hybrid catalysts. *Journal of Environmental Chemical Engineering*, 2021. **9**(2): p. 105091.
165. Papić, S., et al., Decolourization and mineralization of commercial reactive dyes by using homogeneous and heterogeneous Fenton and UV/Fenton processes. *Journal of Hazardous Materials*, 2009. **164**(2–3): p. 1137–1145.

4 Nanomaterials

NANOPARTICLES

Iron-based nanoparticles have shown promise in heterogeneous Fenton processes. They exhibit superb reactivity and specificity towards pollutants due to their small size and high surface-to-volume ratio. This allows for the efficient oxidation of contaminants in wastewater. Further, ion-based nanoparticles have a redox potential that allows for the reduction of hydrogen peroxide to form hydroxyl radicals ($^{\bullet}OH$), which are strong oxidizing agents that can effectively degrade pollutants in wastewater. Their solubility is low under anaerobic conditions, making them suitable for anaerobic environments. Nanoparticles are relatively low cost compared to other advanced oxidation processes, such as photocatalysis, and can be easily synthesized. They can also be recovered and reused, which makes them a more sustainable and cost-effective option for wastewater treatment. Examples of iron minerals and iron-based materials in the nanoparticulate form include magnetite (Fe_3O_4), goethite (α-FeOOH), hematite (Fe_2O_3), ferrihydrite ($Fe_5HO_8{\cdot}4H_2O$), akaganeite (β-FeOOH), greigite (Fe_3S_4), or zero-valent iron nanoparticles (nZVI). These nanoparticulate forms of iron minerals and iron (hydr)oxides can have unique properties and reactivity compared to their bulk counterparts, making them attractive for various applications, including Fenton processes [1]. Examples of the nanostructures used in the Fenton process are shown in Table 4.1.

Iron-based nanoparticles have shown promise in heterogeneous Fenton processes. They exhibit superb reactivity and specificity towards pollutants due to their small size and high surface-to-volume ratio. This allows for the efficient oxidation of contaminants in wastewater. Further, ion-based nanoparticles have a redox potential that allows for the reduction of hydrogen peroxide to form hydroxyl radicals ($^{\bullet}OH$), which are strong oxidizing agents that can effectively degrade pollutants in wastewater. Their solubility is low under anaerobic conditions, making them suitable for anaerobic environments. Nanoparticles are relatively low-cost compared to other advanced oxidation processes, such as photocatalysis, and can be easily synthesized. They can also be recovered and reused, which makes them a more sustainable and cost-effective option for wastewater treatment. Examples of iron minerals and iron-based materials in the nanoparticulate form include magnetite (Fe_3O_4), goethite (α-FeOOH), hematite (Fe_2O_3), ferrihydrite ($Fe_5HO_8{\cdot}4H_2O$), akaganeite (β-FeOOH), greigite (Fe_3S_4), or zero-valent iron nanoparticles (nZVI). These nanoparticulate forms of iron minerals and iron (hydr)oxides can have unique properties and reactivity compared to their bulk counterparts, making them attractive for various applications, including Fenton processes [15,16].

Magnetite is a naturally occurring iron oxide with magnetic properties that is commonly used in various applications, including environmental remediation and

DOI: 10.1201/9781003364085-4

TABLE 4.1
Examples of the Nanostructures Used in the Fenton Process (Self-prepared)

Nanostructure	Size	Shape	Reference
Fe-G graphene-modified iron sludge	–	curled sheet	[2]
Cu-doped Fe/Fe_2O_3	20–100 nm	abundant spherical nanoparticles	[3]
Iron-modified rectangle	500 nm	–	[4]
Fe-SBA-15 nanocomposites	–	–	[5]
Magnetite nanostructure	20–40, 100–200 nm	–	[6]
a-Fe2O3 nanograins	40–280 nm	grains	[7]
Nanophotocatalyst modified by graphite electrode	8–15, 3–5 nm	nanotubes	[8]
$FeVO_4$ with CeO_2	–	nanotubes	[9]
$BiFeO_3$ nanostructures	100–150 nm	–	[10]
$AlSi_2Fe_6$	215.7	–	[11]
Nanometric magnetite (Fe_3O_4)	17–25 nm	ball-milled	[12]
Fe_3O_4	75 nm	spherical particles with rough surface	[13]
Chlorpheniramine	50 ± 20 nm	powder	[14]

biomedical imaging. Due to their unique properties, magnetite nanoparticles have been explored as an alternative to traditional iron-based catalysts in Fenton processes. The magnetic nature of magnetite nanoparticles allows for easy separation from the reaction mixture using an external magnetic field, which can reduce the cost and complexity of the separation process. Additionally, the high surface area-to-volume ratio of nanoparticles can enhance the catalytic activity of the magnetite catalyst, leading to improved Fenton process efficiency [17].

Recent studies have reported the successful application of magnetite nanoparticles in Fenton processes for the degradation of various organic contaminants in wastewater. For example, magnetite nanoparticles were used as catalysts in the Fenton process to degrade methyl orange and Congo red dyes, resulting in high degradation efficiencies. Additionally, magnetite nanoparticles were found to be effective catalysts in the Fenton-like process for the removal of organic pollutants from landfill leachate [18]. The use of magnetite nanoparticles in Fenton processes has also been investigated in combination with other advanced oxidation processes, such as photocatalysis and electro-Fenton [19]. In these combined processes, magnetite nanoparticles are used as a catalyst to generate hydroxyl radicals through the Fenton reaction, while other mechanisms generate additional reactive species. This synergistic effect has been shown to enhance the degradation efficiency of various organic contaminants in wastewater. Altogether, data shows that magnetite nanoparticles have shown great potential as catalysts in Fenton processes for degrading organic contaminants in wastewater. Their magnetic nature and high surface-area-to-volume ratio make them efficient catalysts with easy separation. At the same time, their unique properties also make them promising candidates for combined advanced oxidation processes.

Goethite, an iron hydroxide mineral, also can form nanoparticles of needle-like whiskers and has applications in catalysis, adsorption, and water treatment. Its unique crystal structure and surface chemistry make it an effective catalyst for degrading various organic pollutants in wastewater [20]. Goethite nanoparticles can be synthesized through multiple methods, such as hydrothermal synthesis, sol-gel synthesis, and coprecipitation [21]. In the Fenton processes, goethite nanoparticles can act as a heterogeneous catalyst, where the pollutants in the wastewater can adsorb onto the surface of the goethite particles, facilitating their oxidation by the hydroxyl radicals generated from the Fenton reaction. However, the efficiency of goethite as a catalyst in Fenton processes can be influenced by various factors, such as the size, shape, and surface area of the nanoparticles, as well as the pH and temperature of the reaction. Recent studies have shown that goethite nanoparticles can be modified with other metal ions, such as Cu^{2+} and Ni^{2+}, to enhance their catalytic activity in Fenton processes [22]. These modified goethite nanoparticles have been found to exhibit improved degradation efficiency towards various organic pollutants, including dyes and pharmaceuticals, in wastewater treatment.

Hematite is an iron oxide mineral with a reddish-brown color and is widely used in pigments and in environmental applications such as arsenic removal from water. Hematite nanoparticles have been investigated for their potential use as catalysts in Fenton-like reactions for wastewater treatment [23]. Research has shown that the addition of hematite nanoparticles can enhance the degradation of organic contaminants in wastewater through the Fenton-like reaction, by generating hydroxyl radicals from H_2O_2 in the presence of Fe^{2+} ions. The high surface area and porosity of hematite nanoparticles can also enhance their catalytic activity, allowing for more efficient and effective degradation of contaminants. Additionally, the stability and reusability of hematite nanoparticles make them a promising candidate for practical applications in wastewater treatment. Furthermore, recent studies have explored the use of modified hematite nanoparticles, such as those coated with graphene oxide or doped with other metals, to enhance their catalytic properties further and improve their performance in Fenton processes [24].

Ferrihydrite is a poorly crystalline iron oxide/hydroxide mineral with a high surface area commonly found in soils and sediments. Ferrihydrite nanoparticles have a unique crystalline structure with a high surface area, making them promising materials for advancing Fenton processes. The high surface area allows for more active sites for the catalytic reaction to take place, leading to increased efficiency in removing contaminants from wastewater [25]. In addition, ferrihydrite nanoparticles have been found to have good stability and reusability in Fenton processes, making them a cost-effective solution for wastewater treatment. Researchers have also explored modifying the surface chemistry of ferrihydrite nanoparticles to enhance their catalytic activity and selectivity for specific contaminants. These nanoparticles can potentially be effective catalysts for Fenton processes in wastewater treatment due to their unique structure, high surface area, and stability.

Akaganeite forms a corrosion product in acidic environments and has potential applications in water treatment and environmental remediation. Akaganeite is an iron hydroxide mineral with a unique crystal structure and surface reactivity compared to other iron oxides/hydroxides. Its high specific surface area, acidic properties, and ability to undergo phase transformations make it a promising material for Fenton processes [26]. The use of akaganeite nanoparticles in Fenton reactions has been shown to effectively degrade organic contaminants, such as bisphenol A and phenol, in wastewater treatment. In addition, akaganeite nanoparticles can be easily synthesized and have shown to be stable under a wide range of pH and temperature conditions, which makes them a suitable candidate for practical applications. However, more research is needed to optimize their performance and understand the underlying reaction mechanisms involved in Fenton processes with akaganeite nanoparticles. Their unique properties offer new opportunities for advancing the effectiveness and efficiency of Fenton processes in water treatment and environmental remediation.

Greigite (Fe_3S_4) is a magnetic iron sulfide mineral with considerable potential in environmental remediation and bioremediation. Greigite is a magnetic iron sulfide mineral that has been explored for its potential applications in environmental remediation and bioremediation [27]. It has a cubic crystal structure and is stable under various environmental conditions. Due to their magnetic properties, greigite nanoparticles can be manipulated and controlled using magnetic fields, which makes them useful for the targeted delivery and separation of pollutants in contaminated environments. In Fenton processes, greigite nanoparticles have been studied as potential catalysts for the degradation of organic pollutants. The greigite nanoparticles' high surface area and reactivity make them effective catalysts for the Fenton reaction. They have been shown to enhance the rate and efficiency of pollutant degradation. Additionally, the magnetic properties of greigite nanoparticles make them easy to separate from the treated wastewater, which could reduce the cost and complexity of the treatment process. However, further research is needed to fully understand their properties and optimize their use in Fenton processes.

Zero-valent iron nanoparticles (nZVI) are promising in heterogeneous Fenton processes. The nZVI has shown promising results for the removal of organic pollutants from wastewater, and ongoing research is focused on optimizing the synthesis, characterization, and application of nZVI in wastewater treatment [28]. Several types of zero-valent iron nanoparticles have been used in heterogeneous Fenton processes for wastewater treatment.

These include:

- Bare nZVI: This is the most common type of nZVI used in Fenton processes, where the iron nanoparticles are synthesized without any surface modification or coating [29].
- Coated nZVI: Iron nanoparticles can be coated with various materials to improve their stability, reactivity, and selectivity. For example, silica-coated nZVI (nZVI@SiO_2) and starch-coated nZVI (nZVI@starch).
- Supported nZVI: Iron nanoparticles can be supported on various materials to enhance their dispersion and stability. This can be nZVI supported on activated carbon (nZVI@AC) and nZVI supported on clay (nZVI@clay) have been used in Fenton processes.

- Composite nZVI: Iron nanoparticles can be incorporated into various composites to improve their performance and functionality. For example, nZVI/polymer composites, nZVI/activated carbon composites, and nZVI/iron oxide composites.

BARE ZVI NANOPARTICLES

Bare nZVI refers to zero-valent iron nanoparticles that are synthesized without any surface modification or coating. Bare nZVI is the most common type of nZVI used in Fenton processes for wastewater treatment. Under inert conditions, nZVI can be prepared by reducing iron salts in the presence of a reducing agent, such as sodium borohydride or hydrazine.

They have several advantages for use in Fenton processes. Firstly, it has a high reactivity due to its small size and high surface area, which allows for the efficient oxidation of pollutants in wastewater. Secondly, it is easy to prepare and can be synthesized using low-cost materials. Thirdly, bare nZVI can be used in various environmental conditions, including anaerobic conditions, making it suitable for use in various wastewater treatment applications [30].

However, there are also some challenges associated with the use of bare nZVI in Fenton processes. One of the main challenges is that the nanoparticles tend to aggregate and form larger particles, which can reduce their reactivity and hinder their ability to degrade pollutants efficiently. Additionally, bare nZVI is prone to oxidation and corrosion, which can reduce its effectiveness and lead to the formation of iron sludge [30].

Researchers have explored various surface modifications and coatings for nZVI to overcome these challenges to improve their stability and reactivity. These include coatings such as silica, starch, and polymers, and supports such as activated carbon and clay. However, despite the development of coated and supported nZVI, bare nZVI remains a popular and effective option for Fenton processes in wastewater treatment.

COATED nZVI

The coating on nZVI can be associated with various materials to improve their stability, reactivity, and selectivity in Fenton processes for wastewater treatment. Coatings can provide a protective layer on the surface of the nanoparticles, preventing oxidation, aggregation, and corrosion. Coatings can also improve the dispersibility of nZVI in water, making it easier to mix with other reagents and enhancing its reactivity.

Silica-coated nZVI (nZVI@SiO_2) is a commonly used coated nZVI in Fenton processes. Silica is an inert material that is resistant to oxidation, and its coating can provide a protective layer on the surface of the iron nanoparticles. The coating can also enhance the dispersibility of nZVI in water, allowing for better contact between the nanoparticles and the pollutants. Silica-coated nZVI has been shown to have improved stability and reactivity in Fenton processes, leading to more effective degradation of pollutants in wastewater [31].

Starch-coated nZVI (nZVI@starch) is another type of coated nZVI that has been used in Fenton processes. Starch is a natural polysaccharide that can be used as a stabilizing agent for nZVI. The coating can improve the dispersibility of nZVI in water and enhance its reactivity by providing a source of hydrogen peroxide through the decomposition of the starch. Starch-coated nZVI has been shown to be effective in the removal of organic pollutants, such as dyes and phenols, from wastewater [32].

Other materials that have been used as coatings for nZVI in Fenton processes include polymers, such as polyvinyl alcohol (PVA), and graphene oxide (GO) [33]. Polymer coatings can improve the stability and dispersibility of nZVI in water, while GO coatings can enhance the reactivity of nZVI through the generation of hydroxyl radicals. The choice of coating material depends on the specific application and the desired properties of the coated nZVI.

SUPPORTED nZVI

Supported nZVI refers to zero-valent iron nanoparticles that are supported on various materials to enhance their dispersion and stability in Fenton processes for wastewater treatment. The support material can improve the dispersibility of nZVI in water, prevent aggregation, and provide a protective layer to prevent oxidation and corrosion. The choice of support material depends on the specific application and the desired properties of the supported nZVI.

Activated carbon (AC) is a commonly used support material for nZVI in Fenton processes. AC has a high surface area and porosity, allowing for the adsorption of pollutants and the dispersion of nZVI in water. nZVI supported on AC (nZVI@AC) has been shown to be effective in the removal of various pollutants, including organic dyes, pharmaceuticals, and heavy metals from wastewater [34,35].

Clay is another material that can be used as a support for nZVI in Fenton processes. Clay has a layered structure and a high surface area, allowing for the adsorption of pollutants and the dispersion of nZVI in water. nZVI supported on clay (nZVI@clay) has been shown to have improved stability and reactivity in Fenton processes, leading to more effective removal of pollutants from wastewater [36].

Other materials that have been used as supports for nZVI in Fenton processes include magnetite, zeolites, and chitosan. Magnetite can enhance the magnetic properties of nZVI, allowing for easier separation and recovery of the nanoparticles [37]. Zeolites and chitosan can provide a source of hydrogen peroxide through the release of iron ions, enhancing the reactivity of nZVI in the Fenton processes [38].

Supported nZVI has several advantages over bare nZVI, including improved stability, dispersion, and reactivity. However, the use of support materials can also introduce additional complexity and cost to the synthesis and application of nZVI in Fenton processes. The choice of support material and synthesis method must be carefully considered to optimize the effectiveness and efficiency of supported nZVI in wastewater treatment.

OTHER COMPOSITES

Incorporating nZVI into various matrices enhances the performance and functionality of Fenton processes for wastewater treatment. The composites can provide additional functionalities such as improved stability, selectivity, and removal efficiency, or can enhance the properties of nZVI such as magnetic properties or catalytic activity. Polymer composites are one type of composite nZVI that has been used in Fenton processes. nZVI can be incorporated into polymer matrices such as polyvinyl alcohol, polyethylene glycol, and polyurethane to improve the dispersibility of the nanoparticles and prevent aggregation. Polymer composites have been shown to have improved stability and reactivity in Fenton processes compared to bare nZVI [39].

The nZVI can be further combined with semiconducting materials. Semiconductors have been identified as potential catalysts due to their high optical activity and low cost [40]. One such material is TiO_2, which has been used to form a nanoparticle layer on iron-based catalysts such as Fe_2O_3 nanograin [7]. The effectiveness of using TiO_2 nanostructures as catalysts depends on the amount of remaining compounds present. Studies have shown that the antimicrobial properties of the catalyst increased 2.5 times after using H_2O_2. In another study, researchers used a novel method to synthesize a catalyst for wastewater treatment. Specifically, they used porous $FeVO_4$ nanotubes decorated on CeO_2 nanotubes ($FeVO_4$ with CeO_2) and applied ultrasonic (US), ultraviolet (UV), and binary US/UV radiation [9]. This unique approach led to the formation of highly stable and efficient sonophotocatalysts, which retained their effectiveness even at high pH levels. The resulting catalyst was then used in a special three-way mechanism of Fenton's experiments, which relies on the trapped active species and the calculated energy of the forbidden gap. The study demonstrated the potential of this method in producing highly effective catalysts for wastewater treatment, which could have significant implications in the field of environmental remediation.

Other semiconductors that can be used as catalysts with iron include CdS [40,41], vanadate, $BiVO_4$ [42,43], noble metals (such as Ag), and metal-free semiconductors such as g-C_3N_4 [44]. In Fenton processes, nanoparticles of various sizes are used. For instance, $BiFeO_3$ having 100–150 nm [10], 20–40 and 100–200 nm magnetite-based nanostructures synthesized with plasma [6], magnetite nanostructures with dimensions of 22.7 and 15.1 nm [45], or 17–25 nm nanometric ball-milled magnetite [12]. Garrido-Ramirez *et al.* considered nanostructured allophane clays supported on iron oxide ($AlSi_2Fe_6$) as well as the graphite-modified glassy carbon (GC) and $AlSi_2Fe_6$ [11]. However, the most popular semiconductors are MOF, TiO_2, g-C_3N_4, and bismuth-like materials due to their potential interactions with various types of iron and optical properties [40]. Researchers have used various nanomaterials such as $BiFeO_3$, magnetite, and pyrite nanostructures, as well as nanostructured allophane clays supported on iron oxide and graphite-modified glassy carbon, to investigate the potential of combining different types of nanocatalysts. These studies suggest that combining different nanocatalysts can lead to synergistic effects that enhance the efficiency of the Fenton process for wastewater treatment.

PROCESS PARAMETERS

Some of the important process parameters in Fenton processes include pH, temperature, hydrogen peroxide (H_2O_2) concentration, nano-iron catalyst concentration, reaction time, or presence of contaminants. The pH of the solution can have a significant impact on the Fenton reaction rate and efficiency. Temperature can affect the rate of Fenton reactions, as higher temperatures can lead to faster reactions but can also increase the risk of catalyst deactivation. The concentration of H_2O_2 and iron catalyst concentration can affect the rate of Fenton reactions such as faster reactions but can also increase the risk of unwanted side reactions or catalyst deactivation. The presence of contaminants in the wastewater can affect the Fenton reaction rate and efficiency, as they may compete with the target contaminants for the available H_2O_2 and catalyst [46].

The pH of the solution is an important parameter in Fenton processes that use iron-based nanomaterials as catalysts. The optimal pH range for the Fenton process is typically between 2.5 and 4, which corresponds to the acidic range. At this pH range, the solubility of Fe^{3+} ions is higher, which leads to a higher concentration of Fe^{2+} ions in the presence of H_2O_2. The same relates to nanoparticulate iron-based catalysts. The increased concentration of Fe^{2+} ions enhances the Fenton reaction rate, resulting in more efficient degradation of pollutants in the wastewater. However, it is important to note that the pH can also influence the stability and reactivity of iron-based nanomaterials. For example, at low pH values, bare nZVI may experience rapid oxidation and passivation, leading to a decrease in its reactivity. Therefore, the pH of the solution should be carefully controlled and optimized to ensure that the iron-based nanomaterials exhibit the desired catalytic activity and stability [46].

Temperature is another important parameter that can affect the Fenton reaction rate in the context of iron-based nanomaterials used as catalysts. Higher temperatures can lead to faster reactions due to the increased kinetic energy of the reactants, but can also increase the risk of catalyst deactivation due to thermal degradation. For instance, bare nZVI can undergo thermal oxidation at high temperatures, resulting in a loss of its reactivity towards H_2O_2. Similarly, coated or supported nZVI may experience changes in their surface properties or structure at high temperatures, leading to a decrease in their catalytic activity. Therefore, it is important to carefully control the temperature during the Fenton processes to ensure that the iron-based nanomaterials exhibit the desired catalytic activity and stability. The optimal temperature range may vary depending on the type of nanomaterial used, the composition of the wastewater, and other process parameters. However, a temperature range of 20–40°C is typically used for Fenton processes employing iron-based nanomaterials as catalysts [47].

Reaction time affects Fenton reactions using iron-based nanomaterials. The duration of the Fenton reaction can affect the overall efficiency of the process by influencing the extent of pollutant degradation and the consumption of hydrogen peroxide. Shorter reaction times may not allow for the complete degradation of pollutants, resulting in incomplete treatment. On the other hand, longer reaction times may lead to excessive consumption of hydrogen peroxide and the formation of unwanted byproducts, which can decrease the efficiency of the process. Additionally, the optimal reaction time can depend on the specific type of iron-based nanomaterial being used, as well as other process parameters such as pH and hydrogen peroxide concentration. Therefore, the reaction time should be carefully optimized to balance the desired extent of pollutant degradation with the consumption of hydrogen peroxide and the formation of unwanted by-products [48].

The concentration of iron catalyst can affect the rate of the reaction, as higher concentrations can lead to faster reactions due to the availability of more active sites for the reaction. However, higher concentrations of iron catalysts can also increase the risk of catalyst deactivation due to the formation of iron oxide or other iron-containing species that can reduce the catalytic activity of the iron-based nanomaterials. The use of nano-iron can potentially shorten the Fenton reaction time compared to traditional Fenton processes using bulk iron or other iron-based catalysts. This is because nano-iron has a higher surface area to volume ratio, which can increase its reactivity and efficiency in the Fenton reaction. The increased surface area of nano-iron can provide more active sites for the reaction to occur, leading to faster reaction rates and shorter reaction times. Additionally, the small size of nano-iron particles can improve their dispersibility and accessibility to pollutants, further enhancing the efficiency of the process. However, it is important to note that the optimal reaction time can depend on various factors such as the specific type of nano-iron, pH, hydrogen peroxide concentration, and pollutant type and concentration. Additionally, the optimal concentration of iron catalyst can depend on the specific type of iron-based nanomaterial being used. For example, bare nZVI particles typically require higher concentrations of iron catalyst to achieve the effective catalytic activity, while coated or supported nZVI particles may require lower concentrations due to their improved stability and reactivity. Thus, the concentration of iron catalyst should be carefully optimized to balance the desired reaction rate and the risk of catalyst deactivation [49].

The H_2O_2 plays a crucial role in Fenton reactions by providing the hydroxyl radical ($^\bullet OH$) that is responsible for the oxidation of organic pollutants. The concentration of H_2O_2 can affect the reaction rate, with higher H_2O_2 concentrations generally leading to faster reactions. However, excessively high H_2O_2 concentrations can lead to unwanted side reactions, such as the generation of superoxide radicals ($^\bullet O_2^-$) and hydroperoxyl radicals ($HO_2^\bullet$) that can compete with the formation of $^\bullet OH$ radicals and reduce the efficiency of the Fenton reaction. Therefore, it is important to carefully optimize the H_2O_2 concentration in Fenton processes in relation to iron-based nanoparticle usage, to ensure that the desired level of $^\bullet OH$ radicals is generated without the production of unwanted side products. The optimal H_2O_2 concentration may depend on various factors, such as the type of iron-based nanomaterial used, the type and concentration of pollutants in the wastewater, and other process parameters. In general, the H_2O_2 concentration is typically maintained in the range of 1–10 mM in Fenton processes employing iron-based nanomaterials as catalysts. However, the optimal concentration may vary depending on the specific application and the desired reaction rate and selectivity [50].

The presence of other contaminants in the wastewater can have a significant impact on the efficiency of Fenton processes that use iron-based nanomaterials. For example, organic compounds, such as humic acid, can react with H_2O_2 and consume it before they can react with the target contaminant. This can reduce the efficiency of the Fenton process and result in incomplete removal of the target contaminant. Similarly, the presence of heavy metals in wastewater can affect the performance of iron-based catalysts, as they may bind to the active sites on the catalyst surface and reduce their reactivity. The presence of chloride ions can result in the oxidation of the catalyst and the formation of iron chloride, which reduces the reactivity of the catalyst. Therefore, it is crucial to evaluate the impact of other contaminants on the efficiency of Fenton processes using iron-based nanomaterials and optimize the process conditions accordingly [51].

In heterogeneous Fenton processes, catalyst properties such as structure, size, and application are often adjusted to enhance their efficiency. Nanostructured coatings and materials with intermediate sizes between molecular and microscopic structures can also be used as catalysts. Nanostructures with various sizes and shapes influence the process. Iron nanoparticles, nanorods, nanotubes, and nanocrystals are the most commonly used nanocatalysts in HFP due to their electronic, magnetic, optoelectronic, biomaterials, and catalytic properties. The activity of solid catalysts is highly dependent on particle size and structure, as an increased surface area and active sites can enhance catalytic activity. However, the process is associated with some disadvantages, such as the high cost of catalyst generation and toxic organometallic precursors, which can damage cells mechanically. In some cases, nanostructures used as catalysts in heterogeneous Fenton processes have unique antibacterial, photocatalytic, and anti-accumulation properties.

SYNTHESIS

There are several methods that can be used to synthesize nanoparticulate iron-based nanocatalysts for use in Fenton processes, including:

- Coprecipitation: This method involves the simultaneous precipitation of Fe^{2+} and Fe^{3+} salts in the presence of a base or alkali under controlled pH and temperature conditions. The resulting precipitate can be further processed to obtain the desired nanoparticulate catalyst [52]. Coprecipitation is a widely used method for synthesizing nanoparticulate iron-based nanocatalysts for Fenton processes. In this method, Fe^{2+} and Fe^{3+} salts are simultaneously precipitated in the presence of a base or alkali at controlled pH and temperature conditions. The precipitate formed is then further processed to obtain the desired nanoparticulate catalyst. The coprecipitation process typically involves the addition of a base, such as ammonium hydroxide or sodium hydroxide, to a mixture of Fe^{2+} and Fe^{3+} salts, such as $FeCl_2$ and $FeCl_3$, under controlled pH and temperature conditions [53]. The pH and temperature are controlled to ensure the formation of the desired crystal structure and particle size distribution of the resulting nanoparticulate catalyst. After the coprecipitation step, the resulting precipitate is typically washed, dried, and calcined to remove any impurities and improve the catalytic activity of the catalyst. The calcination temperature and time can be adjusted to control the crystallinity and surface properties of the resulting catalyst. Coprecipitation is a relatively simple and cost-effective method for synthesizing iron-based nanoparticulate catalysts for Fenton processes. The resulting catalysts typically have high surface area, high reactivity, and good stability, making them suitable for a wide range of environmental applications.

- Sol-gel method: It is based on the formation of a sol or solution of metal salts, followed by gelation to form a solid material. The resulting gel can then be dried and calcined to obtain the desired nanoparticulate catalyst. The sol-gel method is a widely used method for the synthesis of nanoparticles, including iron-based nanocatalysts for Fenton processes. The process involves several steps, including the formation of a sol, gelation, drying, and calcination. Sol is a stable colloidal suspension of nanoparticles in a liquid medium. In the case of iron-based nanocatalysts, a sol can be formed by dissolving Fe^{2+} and Fe^{3+} salts in a suitable solvent, such as water or alcohol. The sol may also contain a stabilizing agent, such as a surfactant, to prevent particle aggregation. Gelation is the process of converting the sol into a gel or solid material. This can be achieved by various methods, such as adding a cross-linking agent or changing the temperature and pH of the sol. The resulting gel contains a network of nanoparticles. The gel is then dried to remove the solvent and obtain a solid material. This can be done by various methods, such as air-drying or freeze-drying. Calcination is the process of heating the dried material to a high temperature to remove any remaining organic compounds and to induce particle growth and crystallization. The resulting material is a nanoparticulate catalyst that can be used in Fenton processes. The sol-gel method offers several advantages for the synthesis of iron-based nanocatalysts. It allows for precise control over the size and shape of the nanoparticles, as well as the ability to incorporate other materials or dopants into the catalyst. Additionally, the resulting catalyst has a high surface area and porosity, which can enhance its catalytic activity [54].
- Hydrothermal method: This method involves the synthesis of nanoparticles under high pressure and high temperature conditions in a closed vessel. The resulting nanoparticles are usually highly crystalline and have a narrow size distribution. The hydrothermal method is a popular method for synthesizing nanoparticulate iron-based nanocatalysts for use in Fenton processes. The process involves the use of high pressure and high temperature conditions in a closed vessel, which allows for the precise control of the reaction conditions and the resulting nanoparticle properties. In this method, iron salts are typically dissolved in water or another solvent, and a reducing agent or a hydroxide is added to the solution to promote the formation of the desired nanoparticle morphology. The solution is then sealed in a high-pressure reaction vessel and heated to the desired temperature and pressure. During the hydrothermal reaction, the precursors undergo nucleation and growth, leading to the formation of nanoscale particles with controlled size, shape, and composition. The process is often carried out under alkaline conditions to promote the formation of iron oxides or hydroxides, which are commonly used as Fenton catalysts. The resulting nanoparticles are usually highly crystalline and have a narrow size distribution, making them ideal for use as Fenton catalysts. Additionally, the

hydrothermal method can be used to synthesize a wide range of iron-based nanomaterials, including iron oxides, iron hydroxides, and iron sulfides, which can be tailored to specific applications in environmental remediation and water treatment [55,56].

- Reduction method: Reduction of iron salts or oxides using a reducing agent such as sodium borohydride or hydrazine is applied to form nanoparticles. The resulting nanoparticles can be further processed to obtain the desired catalyst. The reduction method is a simple and effective way to prepare iron-based nanoparticles for use in Fenton processes. In this method, iron salts or oxides are typically used as precursors and a reducing agent, such as sodium borohydride or hydrazine, is added to the solution. The reducing agent causes the iron ions to undergo a reduction reaction, resulting in the formation of zero-valent iron (Fe^0) nanoparticles [57]. The size and morphology of the resulting nanoparticles can be controlled by adjusting various reaction parameters such as the concentration of the precursor, the type and concentration of the reducing agent, the reaction temperature, and the reaction time. In general, smaller nanoparticles can be obtained by using higher concentrations of the reducing agent and by reducing the reaction time. The reduction method has several advantages over other synthesis methods, including its simplicity, low cost, and the ability to prepare a wide range of nanoparticle sizes and shapes. However, the method requires careful control of reaction conditions to ensure reproducibility and prevent the formation of unwanted by-products. Once the nanoparticles are synthesized, they can be further processed to obtain the desired catalyst by washing and drying the nanoparticles, and then calcining them at a high temperature. The resulting catalyst can be characterized using various techniques, such as X-ray diffraction (XRD), scanning electron microscopy (SEM), transmission electron microscopy (TEM), and Fourier-transform infrared spectroscopy (FTIR), to ensure that the nanoparticles have the desired size, morphology, and chemical composition.
- Microwave-assisted method: It is based on the use of microwaves to rapidly heat a reaction mixture containing iron salts or precursors. The resulting nanoparticles are typically smaller in size and more uniform than those obtained using conventional methods. The microwave-assisted method is a relatively new technique for the synthesis of iron-based nanoparticles, which has gained popularity due to its several advantages over conventional methods. In this method, a reaction mixture containing iron salts or precursors is subjected to microwave irradiation, which rapidly heats the mixture, leading to the formation of nanoparticles [58]. The microwave-assisted method is an attractive approach because it provides a fast and energy-efficient way of synthesizing nanoparticles. Additionally, this method enables better control over the reaction conditions, such as temperature, pressure, and reaction time, which can result in more uniform nanoparticles with desired properties. The microwave-assisted method can also be used for the synthesis of different types of iron-based nanoparticles, including iron oxide, iron sulfide [59], and iron carbide.

The resulting nanoparticles can have different sizes and shapes depending on the reaction conditions, such as the concentration of reactants and the duration of microwave irradiation. Furthermore, this method can be easily scaled up for the large-scale production of nanoparticles. The microwave-assisted method is a promising approach for the synthesis of iron-based nanoparticles, which can be utilized in various applications, including Fenton processes.

- Electrochemical method: It involves the electrochemical reduction of iron salts or precursors in a suitable electrolyte solution to form nanoparticles. The resulting nanoparticles are typically highly crystalline and have a narrow size distribution. The electrochemical method is a technique that can be used to synthesize nanoparticles of various metals, including iron. The process involves the application of an electric current to a solution containing iron salts or precursors, which leads to the reduction of the metal ions and the formation of nanoparticles. The reaction takes place at the surface of the electrode, which can be made of a variety of materials, such as platinum or carbon. The electrochemical method offers several advantages over other synthesis methods. For example, it allows for precise control of the size and shape of the nanoparticles by adjusting the reaction conditions, such as the current density, temperature, and pH of the electrolyte solution. Additionally, the process can be easily scaled up for industrial production. One variation of the electrochemical method is the template-assisted approach, in which a template is used to control the size and shape of the nanoparticles. The template can be a solid substrate, such as a glass slide or silicon wafer, or a porous membrane. The electrochemical deposition of the metal occurs in the pores of the template, resulting in nanoparticles with uniform size and shape. In summary, the electrochemical method is a powerful technique for synthesizing iron-based nanoparticles for use in Fenton processes, offering precise control over the size and shape of the nanoparticles and the ability to scale up for industrial production [60].

Biological synthesis methods have gained popularity due to their environmentally-friendly nature, as they do not involve the use of toxic substances. Microorganisms play a key role in these methods as their activity leads to the release of metabolites into the solution, which can react with ions or compounds deposited on their surface, leading to the formation of mineral particles. One such method involves the biosynthesis of Fe^{3+} by ferric-reducing bacteria (FRB) through the reduction of Fe^{3+} ions into hydroxide form under anaerobic conditions [61–65]. The Fe^{2+} ions released are then adsorbed with greater excess and are converted to magnetite through sulfate-reducing bacteria (SRB) like *Desulfomonas*. Another biological method is biologically controlled biomineralization (BCM), where microorganisms control the intracellular process of the formation of magnetite crystals. This process is controlled by ligands with stereochemical properties that stimulate the initial layer and induce crystal growth. Magnetotactic bacteria can also be used in biosynthesis [66–68]. Plant extracts like *Camellia sinensis*, *Peumus boldus*, and

Terminalia catappa have also been used as a source of secondary metabolites to mediate the redox and stabilize particles in the synthesis process [69–71].

The process of synthesizing nanoparticles for use as catalysts is a critical step that can greatly influence the resulting properties and performance of the catalyst. There are several factors that should be taken into consideration when selecting a synthesis method, including the desired properties of the catalyst, the scale of synthesis, and the intended application.

The properties of the nanoparticulate catalyst that may be influenced by the synthesis method include the size of the particles, their distribution, morphology, purity, quantity, and process quality with regards to environmental and economic considerations. For instance, a specific size range and distribution may be necessary for optimal catalytic activity, while high purity may be required to avoid unwanted side reactions or toxicity issues. Additionally, the quantity of catalyst produced must be sufficient to meet the demands of the intended application.

The scale of synthesis is another important consideration, as some synthesis methods may be more suitable for large-scale production, while others may be more suitable for small-scale or laboratory-scale synthesis. For example, some methods may require specialized equipment or facilities that are not available on a small scale or may not be economically viable for large-scale production.

Finally, the intended application of the nanoparticulate catalyst should also be taken into account when selecting a synthesis method. Different applications may require different properties, such as high stability, selectivity, or activity, which may be better achieved using specific synthesis methods. Moreover, the synthesis method may also impact the environmental and economic sustainability of the overall process, such as by generating harmful by-products or requiring expensive reagents.

In summary, the choice of synthesis method for nanoparticulate catalysts is a crucial step that should be carefully considered to ensure optimal performance, scalability, and sustainability for the intended application.

NANOCATALYST CHARACTERIZATION

There is a long list of techniques that can be used to characterize iron-based nanocatalysts prior to their use in Fenton processes. These are SEM, TEM, XRD, FTIR, X-ray photoelectron spectroscopy (XPS), Brunauer–Emmett–Teller (BET) analysis, and Zeta potential analysis. These characterization techniques can provide important information about the structure, morphology, surface chemistry, and stability of the nanocatalyst, which can help optimize its performance in Fenton processes [3,72].

SEM is a powerful tool for the characterization of nanocatalysts in Fenton processes. SEM is a type of electron microscopy that uses a beam of high-energy electrons to scan the surface of a sample and generate an image. The electrons interact with the atoms in the sample, producing signals that can be detected and used to create a detailed image of the surface topography and morphology. SEM is particularly useful for determining the morphology and size distribution of the nanocatalyst. The images generated by SEM can reveal the shape and size of the nanoparticles, as well as any surface features or defects. The size distribution of the nanoparticles can be determined by analyzing the SEM images, which can provide information on the average particle size and the size distribution range. In addition to providing information on the morphology and size of the nanocatalyst, SEM can also be used to investigate the surface chemistry of the nanoparticles. By using energy-dispersive X-ray spectroscopy (EDX) or X-ray photoelectron spectroscopy (XPS) in conjunction with SEM, researchers can determine the elemental composition and chemical state of the nanoparticles. This can be useful in identifying any surface modifications or coatings that may be present on the nanoparticles, which can have a significant impact on their reactivity and stability in Fenton processes [9].

TEM is a powerful imaging technique that can provide detailed information about the morphology, size distribution, and crystallinity of nanocatalysts. In TEM, a focused beam of electrons is transmitted through a thin sample, which interacts with the electrons to form an image. TEM can be used to determine the size and shape of individual nanoparticles and their aggregates, as well as the size distribution of the particles in a sample. The resolution of TEM can reach sub-nanometer scale, allowing the observation of the atomic structure of nanoparticles. TEM can also be used to determine the crystal structure of the nanoparticles and their degree of crystallinity, which can affect their catalytic activity. In addition, TEM can be used in combination with other techniques such as energy-dispersive X-ray spectroscopy (EDS) to determine the elemental composition and distribution of the nanocatalysts. This information is important for understanding the chemical nature and activity of the catalysts. Overall, TEM is a valuable tool for characterizing nanocatalysts and understanding their properties and behavior in Fenton processes.

XRD is a technique that is commonly used to identify the crystal structure and crystallinity of nanocatalysts. In XRD analysis, a beam of X-rays is directed at the sample, and the interaction of the X-rays with the atoms in the sample produces a diffraction pattern. The diffraction pattern provides information about the crystal structure of the sample, including the spacing and orientation of the atoms in the crystal lattice. In the context of nano-catalysts for Fenton processes, XRD can be used to identify the crystal structure of the iron-based nanocatalysts, such as bare or coated nZVI, supported nZVI, and composite nZVI. The XRD pattern of the nanocatalyst can help to confirm the presence of iron-based nanoparticles and their crystal structure, which is important for understanding their reactivity and efficiency in Fenton processes. Moreover, XRD can also be used to detect any changes in the crystal structure of the nanocatalyst after being used in the Fenton processes, which may indicate changes in the catalytic activity or stability of the nanocatalyst.

FTIR is a technique used to identify the functional groups present on the surface of nanocatalysts. It works by measuring the absorption or transmission of infrared radiation by the sample. The infrared spectrum produced provides information about the chemical bonds present in the sample. In the case of nanocatalysts for Fenton processes, FTIR can be used to identify functional groups such as hydroxyl groups, carboxylic acid groups, and amine groups on the surface of the catalyst. These functional groups can play a role in the catalytic activity of the nanoparticles by facilitating electron transfer reactions between the catalyst and the reactants. FTIR is a useful technique for characterizing nanocatalysts because it is non-destructive, requires very little sample preparation, and provides detailed information about the chemical composition of the sample. It is often used in combination with other characterization techniques such as SEM and TEM to provide a comprehensive understanding of the properties of the nanocatalyst.

XPS is a surface-sensitive technique that can provide information about the chemical composition and oxidation state of the nanocatalyst. In XPS, X-rays are used to excite electrons from the surface of the nanocatalyst, and the energy of the emitted electrons is measured. The energy of the emitted electrons can be used to identify the elements present in the nanocatalyst and to determine their oxidation state. XPS is particularly useful for studying the surface chemistry of nanocatalysts, as it can provide information about the chemical species present on the surface and the interactions between the nanocatalyst and the surrounding environment. For example, XPS can be used to determine the oxidation state of iron in iron-based nanocatalysts before and after the Fenton process, providing insights into the mechanism of the reaction and the role of the nanocatalyst in the process. Finally, XPS is a powerful tool for the characterization of nanocatalysts, providing valuable information about the chemical composition, oxidation state, and surface chemistry of the material.

BET analysis is a technique used to measure the surface area and porosity of materials, including nanocatalysts. The method is based on the measurement of the adsorption and desorption of a gas, typically nitrogen, on the surface of the material. The amount of gas adsorbed is proportional to the surface area of the material. BET analysis provides information about the specific surface area, pore size distribution, and total pore volume of the nanocatalyst. In the context of nano-catalysts for Fenton processes, BET analysis can be used to determine the available surface area for catalytic reactions. A higher surface area generally means more available sites for reactions to occur, which can increase the efficiency of the Fenton process. Additionally, BET analysis can provide information on the pore size distribution and porosity of the nanocatalyst, which can impact the transport of reactants and products to and from the catalytic sites.

Zeta potential analysis is a technique that can be used to measure the surface charge and stability of nanoparticles. When nanoparticles are dispersed in a liquid medium, they acquire a surface charge due to the adsorption of ions from the medium. The zeta potential is the electric potential difference between the surface of the nanoparticle and the surrounding liquid medium, which indicates the magnitude of the surface charge. Zeta potential analysis involves measuring the movement of the nanoparticles under the influence of an electric field. The magnitude and direction of the movement of the nanoparticles is related to the zeta potential. Nanoparticles with a high zeta potential are more stable in solution because the electrostatic repulsion between particles prevents them from aggregating or settling out of solution. In contrast, nanoparticles with a low zeta potential are more likely to aggregate or form clumps, which can reduce their effectiveness as catalysts. For nano-catalysts used in Fenton processes, zeta potential analysis can be useful in determining the stability of the nanoparticles in the wastewater matrix. The presence of other contaminants in the wastewater can affect the surface charge and stability of the nanoparticles, which can in turn affect their catalytic activity. By measuring the zeta potential of the nanoparticles before and after exposure to wastewater, researchers can evaluate the stability of the nanoparticles and the effectiveness of their surface modification strategies.

Nanocatalysts are complex materials that require a range of characterization techniques to fully understand their structure and properties. These techniques provide information about the morphology, size, composition, surface chemistry, and surface charge of the nanocatalyst, which are all important factors that can influence its performance in Fenton processes [3]. An important aspect of nanocatalysts is to identify and test a plethora of various parameters. For this purpose, some techniques SEM, XRD, XPS, TEM, scanning transmission electron microscopy mapping (STEM), energy-dispersive X-ray spectroscopy (EDX), or atomic force microscopy (AFM) are used, as shown in Figure 4.1.

Altogether, the combination of mentioned techniques provides a comprehensive understanding of the structure and properties of nanocatalysts, which is critical for optimizing their performance in Fenton processes. However, an in-operando techniques can provide real-time monitoring of the Fenton reaction, allowing for a more accurate understanding of the reaction mechanisms and the behavior of the catalyst under actual reaction conditions [73]. This can provide insight into the dynamic changes that occur during the Fenton reaction, such as changes in the surface structure, oxidation state, and composition of the catalyst. For example, in-operando spectroscopic techniques such as in-operando X-ray absorption spectroscopy (XAS) and in-operando Fourier-transform infrared spectroscopy (FTIR) can be used to

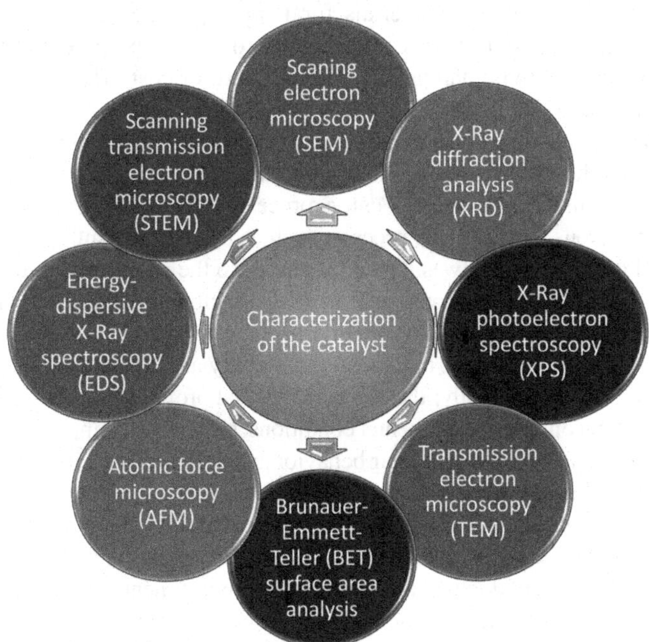

FIGURE 4.1 Techniques used for the characterization of Fenton nano-catalysts.

Source: Own work.

monitor the oxidation state of the catalyst and the evolution of reactive intermediates, such as hydroxyl radicals. In-operando transmission electron microscopy (TEM) and scanning transmission electron microscopy (STEM) can provide information on the changes in the catalyst morphology and structure during the reaction. Finally, in-operando techniques can provide a more complete understanding of the Fenton reaction mechanisms and the behavior of the catalyst, which can help to optimize the process and design more efficient catalysts.

SUMMARY

Iron-based Fenton processes have been demonstrated to be effective in removing a wide range of pollutants from wastewater. However, the use of traditional Fenton processes has limitations such as high operational costs and the need for large amounts of chemicals. Nanomaterials, due to their unique properties, offer great potential in enhancing the efficiency of Fenton processes. Nanomaterials can provide a higher surface area and increased reactivity compared to their bulk counterparts, leading to improved catalytic performance. Additionally, the controlled size and morphology of nanocatalysts can also play a significant role in their catalytic activity. Furthermore, the magnetic properties of some iron-based nanoparticles enable them to be easily separated from the treated water using a magnetic field, which makes them highly suitable for wastewater treatment applications. Despite the potential benefits, there are still challenges that need to be addressed in the use of nanomaterials for Fenton processes. These include the potential release of nanoparticles into the environment, the potential toxicity of certain nanoparticles to human health, and the need for cost-effective and scalable synthesis methods.

The potential release of nanoparticles into the environment is a major concern associated with the use of nanotechnology in various applications, including wastewater treatment using iron-based Fenton processes. The release of nanoparticles can occur at various stages of the process, such as during the synthesis, application, and disposal of the nanocatalysts. Once released into the environment, nanoparticles can interact with various environmental matrices, including water, soil, and air, and potentially affect the ecology and health of organisms. The small size and high surface area of nanoparticles can result in increased reactivity and toxicity compared to their bulk counterparts. Additionally, the physicochemical properties of nanoparticles can change depending on the environmental conditions, such as pH, temperature, and ionic strength, which can further affect their behavior and potential toxicity. Therefore, it is important to ensure that the release of nanoparticles is minimized and that the potential risks associated with their use are carefully evaluated. This includes implementing appropriate safety measures during the synthesis, handling, and disposal of nanoparticles, as well as conducting comprehensive environmental risk assessments to evaluate the potential impacts of their use on the environment [74].

The potential toxicity of certain nanoparticles to human health is an important concern in the field of nanotechnology. Nanoparticles have unique physicochemical properties, including a high surface area to volume ratio and increased reactivity, which can make them more toxic than their bulk counterparts. In particular, nanoparticles may have the ability to enter and accumulate in living cells, leading to potentially harmful interactions with cellular components and processes. Studies have shown that some nanoparticles, such as those made of certain metals or metal oxides, can cause oxidative stress and inflammation in cells, which may contribute to a range of adverse health effects. Other potential health concerns include genotoxicity, immunotoxicity, and carcinogenicity. However, it is important to note that not all nanoparticles are equally toxic, and the toxicity of a nanoparticle can depend on factors such as its size, shape, surface chemistry, and the specific biological system it is interacting with. In addition, advances in nanotoxicology research are helping to identify safe exposure levels and to develop strategies for mitigating potential health risks associated with nanoparticle exposure. Therefore, in terms of future directions, it is important to carefully evaluate the potential toxicity of nanoparticles and to implement appropriate safety measures to minimize the risk of harm to the environment and human health. Furthermore, studies could be conducted on the long-term effects of nanoparticles on the environment and human health to ensure their safe and sustainable use in wastewater treatment applications [34].

The cost-effective and scalable synthesis methods are crucial for the industrial-scale production of nanoparticulate iron-based catalysts for Fenton processes. While there are several methods available for the synthesis of these materials, some of them may not be suitable for large-scale production due to high cost or low scalability. Therefore, there is a need for the development of synthesis methods that can produce large quantities of catalysts at low cost. In addition to the cost and scalability, the synthesis methods should also be environmentally friendly, avoiding the use of hazardous chemicals or the generation of toxic waste. This would not only reduce the overall cost of production but also minimize the potential impact on the environment. Another important factor to consider is the reproducibility and consistency of the synthesis methods. The nanoparticulate iron-based catalysts should be produced with a high degree of uniformity in terms of particle size, morphology, and composition to ensure consistent performance in Fenton processes. Therefore, researchers are actively exploring new synthesis methods that are cost-effective, scalable, environmentally friendly, and produce nanoparticles with high reproducibility and consistency. These efforts would help to overcome the challenges associated with the industrial-scale production of nanoparticulate iron-based catalysts for Fenton processes and advance the application of these materials in wastewater treatment.

Research efforts could focus on developing environmentally friendly synthesis methods for nanocatalysts, as well as investigating the potential synergistic effects of combining different types of nanocatalysts. Investigating the potential synergistic effects of combining different types of nanocatalysts involves exploring the possibility of creating a nanocatalyst with enhanced efficiency for Fenton processes. This approach takes advantage of the unique properties of different nanocatalysts to create a composite nanocatalyst with superior catalytic activity, stability, and selectivity. For example, combining iron-based nanocatalysts with other nanomaterials such as carbon nanotubes or graphene oxide can lead to improved catalytic performance due to the synergistic effects between the two materials. Carbon nanotubes and graphene oxide have a high surface area, excellent conductivity, and strong adsorption capabilities, which can enhance the catalytic activity and stability of the iron-based nanocatalyst.

Combining different types of iron-based nanocatalysts with different crystal structures, sizes, and surface properties can lead to improved catalytic performance; for example, combining hematite and magnetite nanoparticles can lead to enhanced Fenton oxidation of pollutants due to the synergistic effects of the two materials. Hematite nanoparticles have a high surface area and strong adsorption capability, while magnetite nanoparticles have magnetic properties that allow for easy separation and recovery. However, investigating the potential synergistic effects of combining different types of nanocatalysts also requires careful consideration of the compatibility and stability of the different materials in the composite nanocatalyst. It is important to ensure that the different nanocatalysts do not interfere with each other's catalytic activity or stability and that they are compatible with the wastewater matrix. Furthermore, optimizing the synthesis process and the ratio of different nanocatalysts is necessary to obtain the maximum catalytic efficiency and stability of the composite nanocatalyst. Overall, investigating the potential synergistic effects of combining different types of nanocatalysts has the potential to significantly enhance the efficiency of iron-based Fenton processes for wastewater treatment and advance the field toward sustainable and cost-effective treatment solutions.

REFERENCES

1. Aghdasinia, H., et al., Pilot plant fluidized-bed reactor for degradation of basic blue 3 in heterogeneous fenton process in the presence of natural magnetite. *Environmental Progress & Sustainable Energy*, 2017. **36**(4): p. 1039–1048.
2. Guo, S., et al., Graphene modified iron sludge derived from homogeneous Fenton process as an efficient heterogeneous Fenton catalyst for degradation of organic pollutants. *Microporous and Mesoporous Materials*, 2017. **238**: p. 62–68.
3. Luo, T., et al., Efficient degradation of tetracycline by heterogeneous electro-Fenton process using Cu-doped Fe@Fe2O3: Mechanism and degradation pathway. *Chemical Engineering Journal*, 2020. **382**: p. 122970.
4. Zhao, X., et al., Removing organic contaminants with bifunctional iron modified rectorite as efficient adsorbent and visible light photo-Fenton catalyst. *J Hazard Mater*, 2012. **215–216**: p. 57–64.
5. Benzaquén, T.B., et al., Heterogeneous Fenton reaction for the treatment of ACE in residual waters of pharmacological origin using Fe-SBA-15 nanocomposites. *Molecular Catalysis*, 2020. **481**: p. 110239.
6. Khataee, A., et al., Preparation of nanostructured magnetite with plasma for degradation of a cationic textile dye by the heterogeneous Fenton process. *Journal of the Taiwan Institute of Chemical Engineers*, 2015. **53**: p. 132–139.
7. Akhavan, O. and R. Azimirad, Photocatalytic property of Fe2O3 nanograin chains coated by TiO2 nanolayer in visible light irradiation. *Applied Catalysis A: General*, 2009. **369**(1): p. 77–82.
8. Khataee, A.R., et al., Combined heterogeneous and homogeneous photodegradation of a dye using immobilized TiO2 nanophotocatalyst and modified graphite electrode with carbon nanotubes. *Journal of Molecular Catalysis A: Chemical*, 2012. **363–364**: p. 58–68.
9. Eshaq, G., et al., Superior performance of FeVO(4)@CeO(2) uniform core-shell nanostructures in heterogeneous Fenton-sonophotocatalytic degradation of 4-nitrophenol. *J Hazard Mater*, 2020. **382**: p. 121059.

10. Luo, W., et al., Efficient removal of organic pollutants with magnetic nanoscaled BiFeO3 as a reusable heterogeneous Fenton-like catalyst. *Environmental Science & Technology*, 2010. **44**(5): p. 1786–1791.
11. Garrido-Ramírez, E.G., et al., Characterization of nanostructured allophane clays and their use as support of iron species in a heterogeneous electro-Fenton system. *Applied Clay Science*, 2013. **86**: p. 153–161.
12. Hassani, A., et al., Preparation of magnetite nanoparticles by high-energy planetary ball mill and its application for ciprofloxacin degradation through heterogeneous Fenton process. *J Environ Manage*, 2018. **211**: p. 53–62.
13. Guo, H., et al., Degradation of chloramphenicol by pulsed discharge plasma with heterogeneous Fenton process using Fe3O4 nanocomposites. *Separation and Purification Technology*, 2020. **253**: p. 117540.
14. Wang, L., et al., Removal of chlorpheniramine in a nanoscale zero-valent iron induced heterogeneous Fenton system: Influencing factors and degradation intermediates. *Chemical Engineering Journal*, 2016. **284**: p. 1058–1067.
15. Wang, J. and J. Tang, Fe-based Fenton-like catalysts for water treatment: Preparation, characterization and modification. *Chemosphere*, 2021. **276**: p. 130177.
16. Taghizadeh, S.-M., et al., New perspectives on iron-based nanostructures. *Processes*, 2020. **8**(9): p. 1128.
17. Minella, M., et al., Photo-Fenton oxidation of phenol with magnetite as iron source. *Applied Catalysis B: Environmental*, 2014. **154–155**: p. 102–109.
18. Hussain, S., et al., Removal of organics from landfill leachate by heterogeneous Fenton-like oxidation over copper-based catalyst. *Catalysts*, 2022. **12**(3): p. 338.
19. Gholizadeh, A.M., et al., Phenazopyridine degradation by electro-Fenton process with magnetite nanoparticles-activated carbon cathode, artificial neural networks modeling. *Journal of Environmental Chemical Engineering*, 2021. **9**(1): p. 104999.
20. Wang, Y., et al., Goethite as an efficient heterogeneous Fenton catalyst for the degradation of methyl orange. *Catalysis Today*, 2015. **252**: p. 107–112.
21. Jaiswal, A., et al., Synthesis, characterization and application of goethite mineral as an adsorbent. *Journal of Environmental Chemical Engineering*, 2013. **1**(3): p. 281–289.
22. Mohapatra, M., et al., Removal of As(V) by Cu(II)-, Ni(II)-, or Co(II)-doped goethite samples. *Journal of Colloid and Interface Science*, 2006. **298**(1): p. 6–12.
23. Lai, J., S. Xuan, and K.C.-F. Leung, Tunable synthesis of hematite structures with nanoscale subunits for the heterogeneous photo-Fenton degradation of azo dyes. *ACS Applied Nano Materials*, 2022. **5**(10): p. 13768–13778.
24. Twinkle, et al., Graphene oxide (GO)/Copper doped Hematite (α-Fe2O3) nanoparticles for organic pollutants degradation applications at room temperature and neutral pH. *Materials Research Express*, 2019. **6**(11): p. 115026.
25. Zhu, Y., et al., Hydrothermal carbons/ferrihydrite heterogeneous Fenton catalysts with low H2O2 consumption and the effect of graphitization degrees. *Chemosphere*, 2022. **287**: p. 131933.
26. Fang, D., et al., Enhanced catalytic performance of β-FeOOH by coupling with single-walled carbon nanotubes in a visible-light-Fenton-like process. *Science and Engineering of Composite Materials*, 2018. **25**(1): p. 9–15.
27. Shi, X., et al., Stoichiometry-controlled synthesis of pyrite and greigite particles for photo-Fenton degradation catalysis. *New Journal of Chemistry*, 2022. **46**(29): p. 14205–14213.
28. Suryawanshi, P.L., et al., *Chapter 27 - Fenton with zero-valent iron nanoparticles (nZVI) processes: Role of nanomaterials*, in Handbook of Nanomaterials for Wastewater Treatment, B. Bhanvase, et al., Editors. 2021, Elsevier. p. 847–866.
29. Pasinszki, T. and M. Krebsz, Synthesis and application of zero-valent iron nanoparticles in water treatment, environmental remediation, catalysis, and their biological effects. *Nanomaterials*, 2020. **10**(5): p. 917.

30. Lefevre, E., et al., A review of the environmental implications of in situ remediation by nanoscale zero valent iron (nZVI): Behavior, transport and impacts on microbial communities. *Science of The Total Environment*, 2016. **565**: p. 889–901.
31. Guan, Z., et al., Application of novel amino-functionalized nzvi@sio2 nanoparticles to enhance anaerobic granular sludge removal of 2,4, 6-Trichlorophenol. *Bioinorganic Chemistry and Applications*, 2015. **2015**: p. 548961.
32. Yang, C., et al., Does soluble starch improve the removal of Cr(VI) by nZVI loaded on biochar? *Ecotoxicology and Environmental Safety*, 2021. **208**: p. 111552.
33. Chen, X., et al., Review on nano zerovalent iron (nZVI): from modification to environmental applications. *IOP Conference Series: Earth and Environmental Science*, 2017. **51**(1): p. 012004.
34. Duarte, F., et al., Treatment of textile effluents by the heterogeneous Fenton process in a continuous packed-bed reactor using Fe/activated carbon as catalyst. *Chemical Engineering Journal*, 2013. **232**: p. 34–41.
35. Stenzel, M.H., Remove organics by activated carbon adsorption. *Chemical Engineering Progress; (United States)*, 1993. **89**: 4.
36. De León, M.A., et al., Application of a montmorillonite clay modified with iron in photo-Fenton process. Comparison with goethite and nZVI. *Environmental Science and Pollution Research*, 2015. **22**(2): p. 864–869.
37. Yang, B., et al., Enhanced heterogeneous Fenton degradation of Methylene Blue by nanoscale zero valent iron (nZVI) assembled on magnetic Fe3O4/reduced graphene oxide. *Journal of Water Process Engineering*, 2015. **5**: p. 101–111.
38. Raji, M., et al., Nano zero-valent iron on activated carbon cloth support as Fenton-like catalyst for efficient color and COD removal from melanoidin wastewater. *Chemosphere*, 2021. **263**: p. 127945.
39. González-Bahamón, L.F., et al., Photo-Fenton degradation of resorcinol mediated by catalysts based on iron species supported on polymers. *Journal of Photochemistry and Photobiology A: Chemistry*, 2011. **217**(1): p. 201–206.
40. Wang, X., et al., Nanostructured semiconductor supported iron catalysts for heterogeneous photo-Fenton oxidation: a review. *Journal of Materials Chemistry A*, 2020. **8**(31): p. 15513–15546.
41. Shi, J., Z. Ai, and L. Zhang, Fe@Fe2O3 core-shell nanowires enhanced Fenton oxidation by accelerating the Fe(III)/Fe(II) cycles. *Water Res*, 2014. **59**: p. 145–153.
42. Huang, S., et al., Constructing magnetic catalysts with in-situ solid-liquid interfacial photo-Fenton-like reaction over Ag3PO4@NiFe2O4 composites. *Applied Catalysis B: Environmental*, 2018. **225**: p. 40–50.
43. Liu, Y., et al., Active magnetic Fe3+-doped BiOBr micromotors as efficient solar photo-fenton catalyst. *Journal of Cleaner Production*, 2020. **252**: p. 119573.
44. Oh, Y., et al., Divalent Fe atom coordination in two-dimensional microporous graphitic carbon nitride. *ACS Applied Materials & Interfaces*, 2016. **8**(38): p. 25438–25443.
45. Acisli, O., et al., Combination of ultrasonic and Fenton processes in the presence of magnetite nanostructures prepared by high energy planetary ball mill. *Ultrasonics Sonochemistry*, 2017. **34**: p. 754–762.
46. Yu, R.-F., et al., Monitoring of ORP, pH and DO in heterogeneous Fenton oxidation using nZVI as a catalyst for the treatment of azo-dye textile wastewater. *Journal of the Taiwan Institute of Chemical Engineers*, 2014. **45**(3): p. 947–954.
47. Zazo, J.A., et al., Intensification of the Fenton process by increasing the temperature. *Industrial & Engineering Chemistry Research*, 2011. **50**(2): p. 866–870.
48. Mohammad Reza, S., L. Mehdi Ghanbarzadeh, and O. Rabbani, Evaluation of the main parameters affecting the Fenton oxidation process in municipal landfill leachate treatment. *Waste Management & Research*, 2010. **29**(4): p. 397–405.

49. Litter, M.I. and M. Slodowicz, An overview on heterogeneous Fenton and photoFenton reactions using zerovalent iron materials. *Journal of Advanced Oxidation Technologies*, 2017. **20**(1).
50. Wibowo, A., et al., The influence of hydrogen peroxide concentration on catalytic activity of fenton catalyst@bacterial cellulose. *IOP Conference Series: Materials Science and Engineering*, 2019. **509**(1): p. 012020.
51. Chen, Q., et al., Where should Fenton go for the degradation of refractory organic contaminants in wastewater? *Water Research*, 2023. **229**: p. 119479.
52. Jinasan, A., et al., Highly active sustainable ferrocenated iron oxide nanocatalysts for the decolorization of methylene blue. *RSC Advances*, 2015. **5**(40): p. 31324–31328.
53. Zhao, Y., Z. Qiu, and J. Huang, Preparation and analysis of Fe_3O_4 magnetic nanoparticles used as targeted-drug carriers* *Supported by the technology project of Jiangxi Provincial education department and Jiangxi Provincial Science Department. *Chinese Journal of Chemical Engineering*, 2008. **16**(3): p. 451–455.
54. Yoshimura, M. and S. Somiya, Hydrothermal synthesis of crystallized nano-particles of rare earth-doped zirconia and hafnia. *Materials Chemistry and Physics*, 1999. **61**(1): p. 1–8.
55. Lester, E., et al., Reaction engineering: The supercritical water hydrothermal synthesis of nano-particles. *Journal of Supercritical Fluids*, 2006. **37**(2): p. 209–214.
56. Huber, D.L., Synthesis, properties, and applications of iron nanoparticles. *Small*, 2005. **1**(5): p. 482–501.
57. Qin, L., et al., Spherical Zvi/Mn-C bimetallic catalysts for efficient Fenton-like reaction under mild conditions. *Catalysts*, 2022. **12**(4): p. 444.
58. Zanchettin, G., et al., High performance magnetically recoverable Fe3O4 nanocatalysts: fast microwave synthesis and photo-fenton catalysis under visible-light. *Chemical Engineering and Processing - Process Intensification*, 2021. **166**: p. 108438.
59. Yuan, Y., L. Wang, and L. Gao, Nano-sized iron sulfide: structure, synthesis, properties, and biomedical applications. *Frontiers in Chemistry*, 2020. **8**.
60. Bañ Uelos, J., et al., Electrochemically prepared iron-modified activated carbon electrodes for their application in electro-Fenton and photoelectro-Fenton processes. *Journal of The Electrochemical Society*, 2015. **162**: p. E154–E159.
61. Bhatti, H.N., et al., Biocomposite application for the phosphate ions removal in aqueous medium. *Journal of Materials Research and Technology*, 2018. **7**(3): p. 300–307.
62. Kausar, A., et al., Preparation and characterization of chitosan/clay composite for direct Rose FRN dye removal from aqueous media: comparison of linear and non-linear regression methods. *Journal of Materials Research and Technology*, 2019. **8**(1): p. 1161–1174.
63. Remya, V.R., et al., Silver nanoparticles green synthesis: A mini review. *Chemistry International*, 2017. **2**: p. 165–171.
64. Roh, Y., et al., Microbial synthesis and the characterization of metal-substituted magnetites. *Solid State Communications*, 2001. **118**(10): p. 529–534.
65. Mandal, D., et al., The use of microorganisms for the formation of metal nanoparticles and their application. *Appl Microbiol Biotechnol*, 2006. **69**(5): p. 485–492.
66. Bibi, I., et al., Nickel nanoparticle synthesis using Camellia Sinensis as reducing and capping agent: Growth mechanism and photo-catalytic activity evaluation. *Int J Biol Macromol*, 2017. **103**: p. 783–790.
67. Recio Sánchez, G., et al., Leaf extract from the endemic plant Peumus boldus as an effective bioproduct for the green synthesis of silver nanoparticles. *Materials Letters*, 2016. **183**.
68. Muthulakshmi, L., et al., Preparation and properties of cellulose nanocomposite films with in situ generated copper nanoparticles using Terminalia catappa leaf extract. *International Journal of Biological Macromolecules*, 2017. **95**: p. 1064–1071.

69. Truskewycz, A., R. Shukla, and A.S. Ball, Iron nanoparticles synthesized using green tea extracts for the fenton-like degradation of concentrated dye mixtures at elevated temperatures. *Journal of Environmental Chemical Engineering*, 2016. **4**(4, Part A): p. 4409–4417.
70. Kajani, A.A., et al., Anticancer effects of silver nanoparticles encapsulated by Taxus baccata extracts. *Journal of Molecular Liquids*, 2016. **223**: p. 549–556.
71. Bibi, I., et al., Green synthesis of iron oxide nanoparticles using pomegranate seeds extract and photocatalytic activity evaluation for the degradation of textile dye. *Journal of Materials Research and Technology*, 2019. **8**(6): p. 6115–6124.
72. Bury, D., et al., Photocatalytic activity of the oxidation stabilized $Ti_3C_2T_x$ MXene in decomposing methylene blue, bromocresol green and commercial textile dye. *Small Methods*, 2023. **7**: p. 2201252.
73. Yang, Z., et al., Identification and understanding of active sites of non-noble iron-nitrogen-carbon catalysts for oxygen reduction electrocatalysis. *Advanced Functional Materials*. **n/a**(n/a): p. 2215185.
74. Barnes, R.J., et al., The impact of zero-valent iron nanoparticles on a river water bacterial community. *Journal of Hazardous Materials*, 2010. **184**(1): p. 73–80.

5 Application and Efficiency of the Process

The heterogeneous Fenton process is becoming more and more popular to use in environmental protection and engineering. Moreover, it is already entering medicine, creating modern and advanced possibilities for dangerous disease control, especially cancer. Iron, which is necessary for the Fenton and pseudo-Fenton processes, is an important element in the processes taking place in the human body and is important for cell proliferation and growth, or oxygen transport. Iron is found in cells primarily in the form of proteins such as ferritin and hemosiderin. Under special conditions, caused by a high concentration of $HO^•$ radicals, oxidation of unsaturated fatty acids, amino acids, and DNA damage may occur. This process takes place according to the classical Fenton scheme in an acidic lysosome environment. As a result of this process, normal and cancer cells are damaged. Better effects in the fight against diseases may be obtained by using the heterogeneous Fenton process with the use of nanoparticles. However, the amount of hydroxyl radicals produced in the heterogeneous reaction using endogenous H_2O_2 is small and not sufficient to treat cancer cells. However, in recent years, intensive work has been carried out to increase the efficiency of processes with nanomaterials in various processes of modified photo-Fenton and sono-Fenton in cancer cell biological systems [1–4].

However, advanced oxigen processes (AOP) is a demanding and costly process, therefore it is only used for special applications – to remove compounds with high toxicity, and persistence, not susceptible to biodegradation and subject to adsorption and bioaccumulation. We use AOPs to treat wastewater that cannot be effectively treated by biological methods. Scientists noticed a high efficiency of the process during its usage in various types of industries *e.g.*, textile [5–9], pharmaceutical [10], chemical [11], cosmetic [12,13], and during the dewatering of sewage sludge [14] and sewage from the coal gasification process [15–19]. In this type of wastewater containing suspensions and sediments, many organic substances considered hazardous to the environment, can be detected, *i.e.*, dyes, pharmaceuticals, and personal care products. The classical Fenton process is very often used due to its high efficiency, simplicity of use, and low cost. Despite the proven high efficiency of the Fenton process, its use has some disadvantages, such as the need for pH adjustment to initiate the process, providing the appropriate amount of dissolved iron Fe^{2+} in an amount stoichiometric to the number of pollutants, and the proper dose of hydrogen peroxide. The use of iron-based heterocatalysts allows to eliminate most of these disadvantages, which makes its use justified [20].

The effectiveness of the heterocatalytic Fenton process in wastewater treatment depends on both the type of catalyst used and the properties of the wastewater.

DOI: 10.1201/9781003364085-5

Taking into account the properties of the catalyst, iron ions can initiate the Fenton and photo-Fenton processes according to three different physicochemical schemes:

- Homogeneous mechanism involving the effective release of iron ions from the surface of a solid in an acidic environment, followed by a reaction with peroxides,
- Heterogeneous mechanism in which iron cations deposited on (bound to) the catalyst' surface participate in radical formation reactions,
- A semiconductor mechanism in which the charge is photochemically initiated [21,22].

The efficiency of the heterogeneous process according to the homogeneous mechanism depends on the iron ions release rate into the solution. The efficiency of the heterogeneous Fenton process, which takes place according to the heterogeneous mechanism, depends on the properties of the outer surface of the catalyst, *i.e.*, the catalyst particle size distribution, specific surface area, porosity, sorption capacity, and decoration of the catalyst surface. The chemical transformation of compounds takes place in the active sites of the catalyst. The chemical process is preceded by the adsorption of the molecule in the active site, which is limited by mass transfer in solution [23].

The legitimacy of using the heterogeneous Fenton and photo-Fenton process depends also on the type of treated wastewater, *i.e.*, a total load of organic pollutants, the presence of organic ligands, and substances that are radical scavengers. For this reason, we can distinguish two ways of conducting research on the heterogeneous Fenton and photo-Fenton process:

- Research on the decomposition of single model compounds in "synthetic" solutions,
- Research using real wastewater samples [23].

In the first case, we can omit the influence of the matrix on the effectiveness of the process and use solutions of the tested compounds in concentrations significantly exceeding those determined in real samples. In studies of this type, it is necessary to take into account the solubility of the test substance. This type of research primarily concerns substances that are not susceptible to biodegradation, such as dyes or pharmaceuticals, and cosmetic agents. Conducting research on heterogeneous Fenton processes for real wastewater treatment is justified when there is a large share of organic pollutants not susceptible to biochemical decomposition compared to the total content of pollutants. Many researchers decide to undertake this type of research based on the BOD_5/COD ratios determined for the tested wastewater sample or the toxicity tests performed. In studies of this type, it is very difficult to determine the efficiency of the process due to the very diverse qualitative and quantitative composition of the wastewater, which depends on the production profile [23,24].

Economic and social development causes people to pay more and more attention to their health and quality of life. In order to meet the growing expectations of customers, the industry is forced to constantly increase production and improve its products. Therefore, with the rapid increase in industrialization, more and more chemical compounds are produced, which unchanged penetrate into the environment. It is related not only to the emission streams associated with production but also to the use of manufactured goods by the consumer. A good example may be active substances in antibiotics, anti-cancer drugs, antidepressants, etc. These compounds most often end up in the municipal wastewater treatment plant in the stream of pre-treated post-production wastewater or in the stream of domestic wastewater, to which they end up together with urine or faces in the form of metabolites and a certain amount of original molecule. There are cases in that factories do not have their own on-site pre-treatment plants and discharge the entire load directly into the sewage. Moreover, simple biological treatments such as active sludge and biodegradation by microbes are not significantly effective to degrade pharmaceuticals and personal care products and in unchanged form released into the environment and negatively affected. It is therefore highly justified to use AOPs, including the heterogeneous Fenton process and photo-Fenton, for the decomposition of pharmaceuticals and personal care products (PPCP) [25].

The hetorogenenous photo-Fenton process was successfully applied to the decomposition of non-steroidal anti-inflammantory drugs (NSAIDs), antibiotic and antibacterial drugs, and antidepressant drugs. Some common examples of NSAIDs are ibuprofen (IBP), naproxen, diclofenac, piroxicam, and flurbiprofen [26].

Hussain *et al.* achieved almost completely degraded ibuprofen ((RS)-2-(4-(2-methylpropyl)phenyl)propanoic acid, $C_{13}H_{18}O_2$) (IPB) using heterogeneous Fenton process with Cu- and Fe-based catalysts supported over zirconia. Almost complete IBP removal was obtained from the 10 mg/L solutions using catalyst Cu over zirconia (ZrCu) dose 250 mg/L, 3% H_2O_2 dose, at pH 5.0, 70°C after 2 h [27]. Liquid chromatography mass spectrometry (LC-MS) was performed to control the effectiveness of IBP removal and to determine the degradation products during the heterogeneous Fenton process. No toxic metabolites were detected from the oxidation of IBP. However, some IPB oxidation processes using hydroxyl radicals, such as those produced by the heterogeneous Fenton and photo-Fenton processes, have been reported to produce post-reaction mixtures that are more toxic than the initial IBP solution. Due to the complex composition of the mixture after the process, it is not possible to indicate which of the IBP decomposition products has the greatest impact on the increase in toxicity [28]. However, this indicates that the most toxic by-products include dihydroxy ibuprofen, 1-(4-Isobutylphenyl)-1-ethanol and 4-isobutylacetophenone, 2-(4-isobutylphenyl)-5-methylhexan-3-ol, and 1-(4-(1-hydroxy-2methylpropyl)phenyl)ethan-1-one [27,29]. Knowing that chemical compounds, such as IBP, are not completely oxidized to CO_2 and H_2O in the heterogeneous Fenton process, to assess the effectiveness of the process, not only the decrease in concentration of the substrate itself is used, but also the degree of mineralization. Mineralization is determined based on the $TOC_{initial}$ value determined in the solution before (t_0) and TOC_{after} after the process (t_{FP}). Hussain *et al.*, in the study of IBP decomposition, obtained a maximum sample mineralization of 53% under optimal conditions using ZrCu as a catalyst. In the same studies, the use of Fe over zirconia (ZrFe) as a catalyst resulted in only 40% mineralization with an IBP removal rate of 97%. Adityosulindro *et al.*, [30] obtained a similar relationship using Fe-zeolite catalyst (of ZSM5 type) for IBP decomposition, *i.e.*, nearly 90% of its removal with maximum mineralization reaching 27%, resulting in an oxidant utilization efficiency of about 17%. By examining the decomposition of IBP in an aqueous solution at a concentration of 20 mg/L under the designated optimal conditions (4.8 g/L Fe-ZSM5, 6.4 mM H_2O_2, t = 180 min), 16 decomposition products were identified using high-performance liquid chromatography coupled with high-resolution mass spectrometry (HPLC-HRMS), dominated by hydroxylated ibuprofen adducts. The detected products were formed mainly according to the heterogeneous mechanism with the participation of surface iron forms. Iron leaching from the catalyst was low (0.014 and 0.048 mg/L for 1 and 4.8 g/L Fe-ZSM5, respectively), and the contribution of the homogeneous mechanism to the overall effect was negligible [30]. Ivantes *et al.* [31] achieved an equally high level of IBP mineralization (98–100%) using magnesium ferrite nanoparticles ($MgFe_2O_4$) at doses greater than 0.5 g/L and hydrogen peroxide dose greater than 20 mM. Generally, decomposition intermediates are detected in post-reaction solution, but in this case, it was found that complete mineralization occurred after only 40 min of the process, which was confirmed by the determination of IBP concentrations using HPLC and TOC in the initial solution with a concentration of 10 mg/L IBP and after the process.

Most studies on the degradation of IPB using the heterogeneous Fenton process, taking into account factors such as catalyst properties and dose, H_2O_2 dose, pH, and temperature, are performed with the use of aqueous solutions. However, the conclusion of this type of research is that the method can be successfully used to remove IBP from wastewater. According to the literature, organic and inorganic compounds in wastewater can significantly affect the efficiency of the heterogeneous Fenton process, reducing or increasing its efficiency [30,32]. Adityosulindro *et al.* [30] determined the influence of the matrix on the distribution of IBP by conducting two experiments, where in one case a solution of distilled water (DW) was used, and in another wastewater treated from a municipal treatment plant (WW) was used. Both processes were carried out using the same doses of reagents. The degradation of IBP in WW was much slower and less effective. Only the correction of pH to 3.3 in the WW sample allowed us to obtain a similar effect in both samples. This may be due to a shift in the carbonate balance after acidification and the removal of most of the inorganic carbon (IC). It is well known that carbonates and bicarbonates act as radical scavengers, often reducing the efficiency of the process [30].

The use of the heterogeneous Fenton process to degrade other commonly used non-steroidal anti-inflammatory agents, *i.e.*, naproxen (2-(6-methoxy-2-naphthyl) propionic acid) (NPX) [33,34], diclofenac (2-[2-(2,6-dichloroanilino)phenyl]acetic acid) [35], piroxicam (4-hydroxy-2-methyl-N-(2-pyridinyl)-2H-1,2-benzothiazine-3-carboxamide 1,1-dioxide) [36] also proved to be effective. The use of goethite-montmorillonite nanocomposite, crystal sizes from 2.74 to 45.79 nm, in the optimal dose of 1 g/L, allows for the complete removal of NPX within 60 min. The time of complete decomposition can be significantly shortened by the use of a UV source. However, it should be added that a very high efficiency of the heterogeneous Fenton process was obtained using an aqueous solution of NPX at a concentration of 11.5 mg/L, *i.e.*, unprecedented in the natural environment. This is another scientific report, the results of which clearly show that it is possible to achieve high efficiency in removing non-steroidal anti-inflammatory drugs in cases where their initial concentration is very high. Setifi *et al.* [33] show that the contribution of decomposition by radical mechanisms in the total effect of NPX removal is 68%, while 32% of the NPX was removed from the solution by adsorption. Goethite-montmorillonite nanocomposite has very good sorption properties because the immobilization of goethite nanoparticles on montmorillonite allowed to increase in the specific surface area of the nanomaterial from 17.99 to 183.40 m^2/g [33]. It is possible to completely remove NPX using heterogeneous Fenton processes, but the decomposition is not complete to CO_2 and H_2O. Among the decomposition products of NPX, the most frequently detected compounds are: 2-acetyl-6-methoxynaphthalene, 2-ethenyl-6-methoxynaphthalene, 6-methoxy-2-naphthoic acid, 1-(6-hydroxynaphthalen-2-yl)ethenone, and phthalic acid. Achieving a high degree of mineralization of NPX solutions requires a significant extension of the processing time. Setifi *et al.* [33] obtained a 90% degree of mineralization, expressed as the COD reduction in relation to the COD before the process, after 5 hours, conducting the process with the use of a nanocomposite catalyst at a dose of 1 g/L, H_2O_2 0.005 mol/L, at pH = 4.0, T = 20°C [33].

Diclofenac belongs to organochlorine compounds and this fact is the starting point for research on the use of the heterogeneous Fenton process for its decomposition. It was no difference in the case of the research carried out by Bae *et al.* [37] who assumed that if pyrite can be effectively used to remove organochlorine compounds, *i.e.*, trichlorethylene (TCE) and carbon tetrachloride (CT), then why not use it to decompose diclofenac. The tests were performed using diclofenac at the initial concentration of 0.017 mM and pyrite at doses in the range of 0.5–4 mM, H_2O_2 dose of 1 mM. After adding pyrite, Fe^{2+} ions are released from its surface into the solution until equilibrium is reached. It was found that there is a linear relationship between the pyrite dose and the concentration of Fe^{2+} ions in the solution. There is a synergy between the two processes. The process of releasing iron ions from the surface causes a significant decrease in the pH value, thanks to which equilibrium is reached more quickly. Nearly 14% of iron is released from the pyrite surface, causing the distribution of diclofenac according to the homogeneous mechanism in the total effect is significant [37]. Bae *et al.* [37] pointed out in their work the problem of reducing the solubility of diclofenac with a decrease in pH (precipitation from solution instead of decomposition). The introduction of H_2O_2 into the system not only initiates the heterogeneous Fenton process but additionally lowers the pH. It has been shown that the decomposition of diclofenac occurs very quickly, leading to its complete removal after just 120 s, with a ratio of 11:1 using H_2O_2 to decomposed diclofenac. During the process, as the diclofenac concentration decreases, its decomposition products appear, *i.e.*, 2,6-dichlorophenol, 2-chloroaniline, and 2-chlorophenol. The concentration of these compounds decreases during the process. However, complete mineralization of the sample does not take place in a short time (120 s) [37].

Hetorogenic Fenton and photo-Fenton process can be successfully used to remove other drugs, *i.e.*, ciprofloxacin [38] antibiotics and their common such as chloramphenicol (CAP) [39], ibuprofen [27], components of tetracycline antibiotics [40], and acetaminophen [41].

Antibiotics, as the name suggests, work against living matter (biota). Different types of antibiotics include tetracycline, quinolones, aminoglycosides, sulfonamides, glycopeptides, β-guanamines, macrolides, etc. Their widespread use in the fight against bacterial diseases of animals and humans, as well as in securing the industrial breeding of aquatic plants and animals, caused antibiotics and their metabolites getting into waters and soil pose a huge threat to the environment. They can stay in the environment for a very long time because they are non-biodegradable and have the ability to bioaccumulate [42,43]. Antibiotics and products of their metabolic decomposition by humans and animals end up in the environment together with treated municipal wastewater or are directly excreted. The constant increase in the consumption of antibiotics in the world makes the problem of antibiotics and their metabolites occurrence one of the most important environmental problems. They have the potential to cause microorganism antibiotic resistance in aquatic environments even in very low concentrations [43]. Numerous methods and technologies have been used to treat water and wastewater from antibiotics, *i.e.*, adsorption/bioadsorption, biodegradation, coagulation-sedimentation, membrane separation, ozonation. Most of them have their disadvantages, including insufficient effectiveness of antibiotic removal, formation of toxic intermediates, a large amount of waste generated during the process, and high costs of the process [43,44].

A good solution to the problem of environmental pollution with antibiotics and their metabolites may be the use of the hetrogenic Fenton and pseudo-Fenton process for their degradation. The use of heterogeneous Fenton-like processes for the decomposition of antibiotics has many advantages, including a recyclable catalyst, comparatively low treatment cost, and efficiency that allows the decomposition of pollutants to the level required by law [44].

Cao *et al.* and Fu *et al.* [45,46] conducted independent research on the removal of tetracycline ((4S,6S,12aS)-4-(dimethylamino)-1,4,4a,5,5a,6,11,12a-octahydro-3,6,10,12,12a-pentahydroxy-6-methyl-1,11-dioxonaphthacene-2-carboxamide) (TC) with the use of zero-valent iron (ZVI) [45,46]. TC removal was by adsorption, oxidation, and reduction or oxidation and flocculation. In the case of adsorption and flocculation, we can say that the removal of TC from the solution consisted of the transition of the compound from the aqueous phase to the separated solid phase, which was then removed. TC oxidation was very effective in an acidic medium according to the Fenton-like mechanism [45,46]. At first, only TC and ZVI are present in the solution. Cao *et al.* used 500 mL/min aeration, thanks to which O_2 was effectively introduced [45]. Fu *et al.*, in the second stage of the process, used slow flocculation for 4–30 d in conditions allowing the penetration of O_2 from the air [46]. In both cases, reactions initiate the Fenton process and produce $HO^{\bullet}$, which are very efficient oxidants of organic compounds [45,46].

Cao *et al.* achieved a maximum 85% removal of TC by Fenton-like oxidation at an initial TC concentration of 20 mg/L, ZVI concentration of 0.4 g/L, at pH = 2.5, after 30 min. Based on LC-MS, it was found that intermediate products are formed as a result of hydroxylation of TC by electrophilic addition of a radical to the $HO^{\bullet}$ aromatic ring, N-demethylation process. Subsequent decomposition products have a much simpler structure through N-methyl groups, methyl groups, and hydroxyl groups. Then, the double bonds in the aromatic rings are broken, which is an introduction to the formation of further products by breaking the ring and displacing (detaching) the hydroxyl and methyl groups [45,46]. The participation of the heterocatalytic Fenton process with the use of ZVI without the addition of H_2O_2 requires longer process times and ensures conditions for effective O_2 penetration from the atmosphere (aeration, reaction vessel geometry) [45,46]. Sunlight is a factor that improves the efficiency of the process because using ZVI creates conditions for the formation of Fe-TC complexes [46].

Zhang *et al.* used CeO_2 modified zero-valent nanoiron (modified of nZVI with CeO_2) for TC decomposition. The use of CeO_2 as the nZVI carrier reduces the ability of nZVI to agglomerate and increases the catalytic activity of the composite material due to the synergism of the properties derived from the constituent elements. Based on the characteristics of the material particles using XPS, it was found that the share (ratio) of Ce(III) on the surface of Fe^0/CeO_2 was 45%, while on the surface of pure CeO_2 it was only 30%. Ce^{3+} present on the surface of CeO_2 may accelerate electron transfer and favor the generation of $HO^•$ radicals from H_2O_2. Increased availability of $HO^•$ speeds up the process and increases its efficiency. The highest efficiency of TC decomposition (93%) was obtained using Fe^0/CeO_2 and H_2O_2 in doses of 0.1 g/L and 100 mM, respectively, at pH 3.0 after 60 min. Increasing the catalyst dose from 0.01 to 0.05 g/L significantly affected the efficiency of TC removal (from 42 to 89%). A further increase in the Fe^0/CeO_2 dose led to a slight increase in the efficiency of the process. The TC decomposition using Fe^0/CeO_2 was very efficient, about 93% over a wide range of initial pH (pH 3.0 to 5.8). At pH = 7.0, the reduction in effectiveness was just over 25%. For comparison, tests with nZVI were performed. In this case, the best result was obtained in a narrow pH range (pH = 3.0). The efficiency of the decomposition depended on the pH of the initial solution. Differences in process efficiency amounted to over 40%. Therefore, it may be highly justified to use Fe^0/CeO_2 for TC decomposition due to the high efficiency of TC decomposition in a wide range of initial pH, thanks to the use of an effective catalyst with a large material's specific surface area. In addition, this catalyst can be successfully used several times without a significant reduction in efficiency in subsequent cycles [47].

A lot of attention has been devoted recently to the group of composite nanocatalysts, which are created by depositing iron compounds on reduced graphene oxides (rGO). These types of materials are characterized by a porous structure, lower agglomeration, greater swelling, and higher specific area. Interestingly, the use of this type of catalyst allows to generation of reactive oxygen species (ROS) and initiates the heterogeneous Fenton process without adding H_2O_2 [48,49].

Zhuang *et al.* [49] achieved very good results, they perform a tetracycline decomposition study with the α-FeOOH/rGO catalyst. The experiment was performed by adding 100 mg of nanocatalyst and 60 μL of H_2O_2 to 20 ml of a 100 mg/L tetracycline solution. The authors also performed a comparative study replacing the tetracycline solution with a sample of raw water taken from the Jiangsu River. Both experiments confirmed the high efficiency of the proposed method for the degradation of antibiotics. However, in the case of the raw water sample, this was done indirectly, indicating the degree of contamination removal as TOC removal and giving the concentration of selected antibiotics, including tetracycline, in the water sample [49]. It should be noted that in this type of research, the concentration of the test substance in the model system is much higher than that in environmental samples. The second problem we have to face is the difficulty of interpreting the results when we use the values of indicators such as TOC, COD, or BOD to estimate the removal of micro- and nanopollutants with a given method.

Scaria *et al.* [48] studied the decomposition of TC with the use of magnetite–reduced graphene oxide nanocomposite (MG) in redistilled water solutions using different doses of the catalyst, H_2O_2 in a wide range of initial pH variability. The highest degree of TC reduction was 76.8%. The efficiency of the process increased with the increase of the H_2O_2 dose in the range from 50 to 150 mM. Further increasing the concentration of H_2O_2 resulted in a worsening of the effect. The effectiveness of TC removal changes to a much lesser extent by changing the dose of MG. The optimal ratio of MG (mg/L): H_2O_2 mM was determined to be 2:1. It has been shown that the efficiency of the process increases with the initial concentration of TC. It can be expected that the TC removal efficiency in real samples, in which the TC concentration is much lower, will be significantly lower. The authors themselves estimated that the effectiveness of TC removal with the proposed method with domestic wastewater would be 49% after 150 min.

In addition to pharmaceuticals and personal care products (PPCP), there are still a large number of chemical compounds that are an environmental problem and can be effectively eliminated thanks to the use of the heterogeneous Fenton and photo-Fenton process. Dyes are also included in this group. Their uncontrolled release into the environment is a big social problem because even a small amount of them in the environment causes effects visible to the bare eye. The problem of discharge of untreated or insufficiently treated wastewater containing dyes raises much greater concern for residents than the discharge of other wastewater that is much more harmful but does not have an intense color. The increase in the ecological awareness of the population living in the areas where the factories using dyes operate has resulted in the introduction of legislative changes and continuous monitoring of the elements of the environment to which wastewater is discharged. In the case of wastewater containing dyes, the use of only biological methods for their treatment does not always bring satisfactory results, which is why other methods are being developed, including highly effective oxidation processes, *i.e.*, the heterogeneous Fenton process.

Organic compounds like dyes are used in textile, paint, furniture, cosmetic, and chemical industries. The qualitative composition of the generated wastewater depends very much on the production process. Wastewater from the textile industry is characterized by a high content of organic compounds that are not susceptible to treatment by biological or chemical methods. The poor susceptibility of textile wastewater is determined by the high content of coloring agents, surfactants, and a number of organic compounds, the use of which is aimed at improving the properties of the fabric itself and securing it during transport [8,50,51]. A large amount of water and chemicals are used in the process of dyeing fabrics, with 2 to 50% of the dyes used going to sewage and waste [7]. The situation is different in the case of wastewater from the cosmetics industry, which is generated mainly during the cleaning of equipment for the production and packaging of cosmetics. The contribution of dyes in wastewater compared to the total amount used in production is very small.

Dyes are characterized by greater solubility and diversity – about 10,000 types. The heterogeneous Fenton process has been applied to removed dyes such as acid orange [52–55], methylene blue [56–58], congo red [59], tartrazine [60], reactive black [61], reactive red [62], reactive black [63,64], acid blue [65], basic blue [38], acid red [66], a mixture of azure B and congo red [67], alcian dye blue tetrakis chlorine (methylpyridinium) [68], amaranth [69], and others (see Figure 5.1).

Acid orange (OG), (7-Hydroxy-8-phenylazo-1,3-naphthalenedisulfonic acid disodium salt, $C_{16}H_{10}N_2Na_2O_7S_2$) is an azo dye also known as orange G. There is one azo bond in the molecule (-N=N-) between the aromatic rings, which makes the compound colored. Auxochromes (-OH, $-SO_3^-$) are substituted for the aromatic rings, the presence of which in the molecule results in shifting the chromophore absorption bands and changing their intensity, but also increases the affinity for the solvent, which is water. OG is characterized by intense color, very good solubility in water, and penetration through fabric fibers, which is why it is commonly used in the textile industry at the fabric dyeing stage [70].

Ali *et al.* [55] studied the decomposition of OG using the heterogeneous Fenton process, in which the catalyst was screened waste material from the steel industry (furnace slag and dust). The iron content in the total weight of the waste material was about 90%. Bearing in mind the possibilities of practical use of the test results, an analysis of the catalyst was carried out in terms of the possibility of its separation from the post-reaction mixture by electromagnetic methods and reuse. Based on the

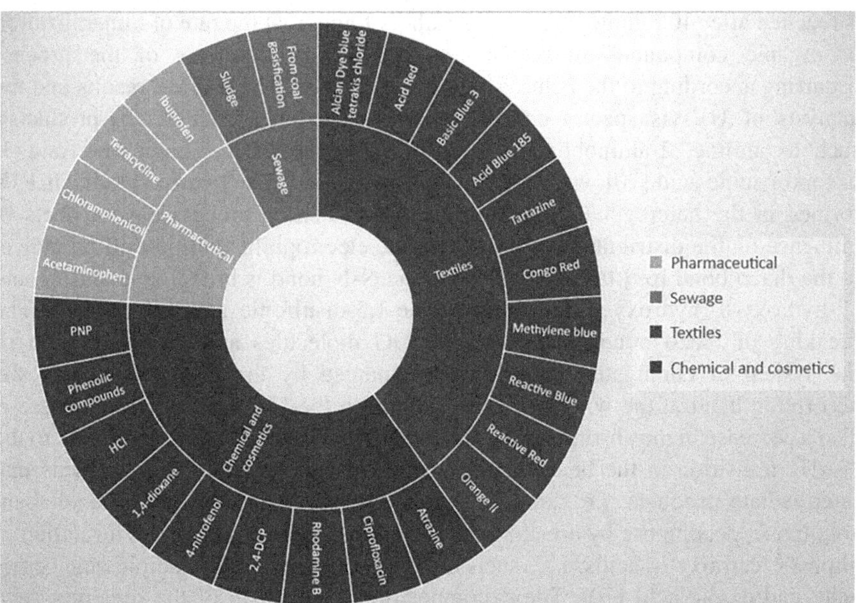

FIGURE 5.1 The types of wastewater decomposed by heterogeneous Fenton process. The figure was created based on approximate value of the results from Table 3.1.

characterization of the material using the XRD method, it was found that it had the desired properties, because magnetite (Fe_3O_4) had the largest share in the mixture of the catalyst used, and hematite (Fe_2O_3) and wuesite ($Fe_{0.925}O$) complemented it. Considering the source of the catalyst, there was no leaching (dissolution) of iron, and the concentration of iron ions in the solution was only 0.15 mg/L. The complete discoloration of the solution with the initial concentration of OG 40 mg/L was obtained after 30 minutes and the use of a catalyst dose of 200 mg/L and 24 mM of H_2O_2, at pH = 2.0 [55]. Research on the use of iron oxides in the process of heterocatalytic decomposition of OG according to the Fenton mechanism was carried out by Wang *et al.* [70]. The proposed solution is very interesting because a three-phase reactor was used in the research, in which the catalyst is in the form of a fluidized bed maintained in the OG aqueous solution thanks to the air supplied in the lower part of the reactor. Thanks to this, there are optimal conditions ensuring very good contact of the catalyst with the substrates in the liquid phase, and additionally, supplying the system with air containing O_2 makes it possible to achieve greater efficiency in the decomposition process. The catalyst was iron oxides, mainly in the form of goethite (a-FeOOH phase). The process was supported by UV light. Almost complete discoloration (92%) of the solution with the initial concentration of OG 50 mg/L was obtained after 60 min and the use of a catalyst dose of 6 and 25 mg/L of H_2O_2, at pH = 3.0. The mineralization of the sample was much slower and the maximum effect of removal was 78% TOC reduction after 3 h. It was indirectly shown that the rate of color decay is determined by the homogeneous mechanism, because iron is washed out of the catalyst very quickly until the maximum concentration of dissolved iron ions (1.4 mg/L) is reached after 40 minutes. The same authors found that the rate of mineralization of organic compounds in the sample depends on the rate of the process occurring according to the heterogeneous mechanism. The GC/MS results and the analysis of UV-vis spectra confirm this. Intermediate products (by-products), such as aniline, 2-aminophenol, 4-nitrophenol, phenol, benzoquinone, and 3-nitropropanoic acid [70], were detected in the solution. The hydroxyl radicals $HO^{\bullet}$ formed in the heterogeneous Fenton and photo-Fenton process have a decisive influence on the distribution of OG. First, the electrophilic addition of the radical to the diazo bond [69] takes place. Then the N-N bond is broken and aniline and 7-hydroxy-8-(hydroxyamino) naphthalene-1,3-disulfonic acid are formed. The breaking of diazo bonds in subsequent OG molecules and the destruction of the system of conjugated bonds is accompanied by the disappearance of the absorption band at the wavelength of 480 nm in the UV-VIS spectrum. There is also a decrease in the absorption bands at 250 and 330 nm, which correspond to the $\Pi \rightarrow \Pi^*$ transition in the benzene and naphthalene rings [69]. This confirms that intermediate products, *i.e.*, aniline, 2-aminophenol, 4-nitrophenol, phenol, benzoquinone, decompose by breaking the aromatic ring, resulting in the formation of aliphatic carboxylic acids, *i.e.*, such as 3-nitropropanoic acid, adipic acid, citric acid, and oxalic acid [70]. The decomposition mechanism of the remaining azo dyes is similar to that described in the example of OG, *i.e.*, electrophilic addition initiating the breaking of the diazo bond, followed by further division of intermediate products.

The use of the heterogeneous Fenton process for the decomposition of dyes including OG is very often determined by economic factors. In order to reduce costs, studies have been started on the degradation of OG using catalysts that can be easily separated and reused without compromising effectiveness. It is well known that the adsorption process on activated carbon is commonly used to remove dyes, but the costs of its use can be high [52]. Therefore, research was undertaken on the use of well-known sorbents with a high proportion of carbon in the composition, immobilized with iron or its oxides in the heterogeneous Fenton and photo-Fenton process. The source of carbon in the prepared catalyst can be by-products from agriculture such as olive stone, stabilized sewage sludge from municipal treatment plants, or specially prepared commercial products such as carbon aerogel. The preparation consists of carbonization in the atmosphere of an inert gas or polymerization, and then immobilization of its surface with iron or its oxides [52,53]. Ramirez *et al.* [52] studied the decomposition of orange II (OGII) in the heterogeneous Fenton process using two different Fe-carbon catalysts. The complete decolorization was obtained at pH 2.0 after 1 h. The rate of OGII decomposition depended on the pH of the solution and decreased with increasing pH. The tests were performed at pH 2.0, 3.0, and 4.0. The maximum, almost 90%, TOC removal was obtained after 4 hours of the process using 0.3 g L^{-1} of catalyst, 6 mM H_2O_2, and pH = 3.0. H_2O_2 doses were determined on the basis of the theoretical H_2O_2 demand for complete oxidation of OGII in solution. By increasing the concentration of H_2O_2 in the range up to 6 mM, an improvement in the efficiency of the dye decomposition was found. Further increasing the concentration did not improve the effect of the decomposition process, and even slightly worsened it. It is worth noting that in the case of the heterogeneous Fenton process with the use of carbon-Fe catalyst, there was no relationship between the applied dose of H_2O_2 and the amount of leached iron, which is commonly observed when using iron or its oxides as a catalyst. The concentration of iron in the solution, leached from the catalyst, changed over time and depended on the pH at which the tests were performed. The most iron was released into the solution at pH = 2.0, and at pH = 4.0 this process was almost completely inhibited. Bearing in mind that at pH 2.0 and 3.0 iron leaching occurs, which may favor the decomposition processes of OGII according to the homogeneous mechanism, it was found that the decomposition according to the heterogeneous mechanism has a much greater share in the total effect.

Tu *et al.* studied the oxidation of acid orange (AO) in the heterogeneous photo-Fenton process using a Fe-carbon catalyst. The source of carbon in the catalyst was properly treated sewage sludge from a municipal wastewater treatment plant. Very good results were achieved, leading to complete discoloration of the solution and 67% mineralization of the sample. The optimal conditions for the dye decomposition were determined experimentally and were as follows: catalyst dose and H_2O_2 1.5 g/L, 15 mM, respectively, at pH = 4.0 and support with visible light (150 W short-arc Xe lamp). Tu *et al.* used relatively higher doses of catalyst and H_2O_2, and the process was supported by light, they achieved very good efficiency of dye decomposition at pH = 4.0 [53]. Can dyes be effectively degraded by $HO^•$ generated in the heterogeneous Fenton process at a pH greater than 4.0? In search of an answer to this question, many researchers have attempted to use a different catalyst that would enable the generation of HO$^\bullet$ in a wider pH range in the same way as in the classical Fenton process. It is well known that the use of copper Cu^+ enables the efficient formation of radicals in the wide-working pH range (pH 3.0–7.0) according to the reactions (Eqs. 5.1 and 5.2) [54].

$$Cu^+ + H_2O_2 \rightarrow Cu^{2+} + HO^\bullet + OH^- \tag{5.1}$$

$$Cu^{2+} + h\nu \rightarrow Cu^+ \tag{5.2}$$

Copper can play a similar role as iron in the Fenton process, which is why it is called a Fenton-like catalyst. However, homogeneous copper is rarely used in decomposition processes because it is toxic to animals inhabiting waters where pretreated wastewater is discharged. However, there is an increase in interest in copper as a catalyst supported on a porous solid substrate [54,69]. Researchers are working on obtaining new ones in which nanoparticles of copper alone or its mixture with iron (iron-copper bimetallic nanoparticle) form composite catalysts when combined with materials such as zeolite, clay, PAN fiber, MCM-41, ZSM-5, and ordered mesoporous carbon materials [69]. The use of this group of heterogeneous catalysts enables the effective decomposition of organic compounds resistant to biodecomposition, including dyes. The heterogeneous pseudo-Fenton process can occur at a higher pH compared to the heterogeneous Fenton. An additional advantage is a significant reduction in the leaching of metals, including heavy metals, *i.e.*, copper, from the catalyst. Thanks to this, it is possible to use the catalyst several times without reducing its catalytic activity [54,69]. Lam *et al.* [54] studied the decomposition of OGII in the pseudo-Fenton process using a heterogeneous catalyst synthesized by supporting copper metal on MCM-41 by chemical vapor deposition. The tests used catalysts at a concentration of 1 g/L, but with different Cu content. The weight share of Cu in the total mass of the prepared catalyst was 0.82% (Cu/M41-1), 1.79% (Cu/M41-2), 2.51% (Cu/M41-3), 3.27 (Cu/M41-4), respectively. During the OGII decomposition process, Cu was leached from the catalyst and its concentration in the solution increased. The concentration of Cu^{2+} ions in the solution after 2 h depends on its content in the catalyst. The lowest concentration of 0.08 mg/L was found with Cu/M41-1, and the highest 0.5 mg/L with Cu/M41-4, respectively. Thus, the concentrations of copper in the solution are below the limit value for wastewater discharged into the environment. The efficiency of the process was defined as the degree of mineralization, or the degree of TOC reduction, with respect to the initial TOC value measured in the 0.3 mM OGII solution. The highest degree of mineralization, 80%, was obtained with the use of 1 g/L of the Cu/M41-4 catalyst with the highest Cu mass fraction, after 2 h of pseudo-Fenton process at pH = 3.0. Difference in mineralization after 2 h depending on the Cu mass fraction in the total mass of the catalyst did not exceed 5%. Replacing Fe with Cu in the M41 composite catalyst allows efficient decomposition of OGII over a wide pH range. Mineralization of OGII at the initial concentration of 0.3 mM and the same doses of the catalyst and H_2O_2 at pH 3.0, 5.5, and 7 are 80, 77, and 70%, respectively [54].

Finding solutions to problems related to the use of heterogeneous Fenton and pseudo-Fenton processes and their modifications for the decomposition of compounds not susceptible to biological degradation, we pose new challenges to researchers. Is it possible to decompose these compounds even more effectively and in a shorter time? The result of this type of research is the development of a new class of heterocatalysts in which more than one metal is present. Wang *et al.* [69] studied the distribution of OGII using iron–copper bimetallic nanoparticles supported on hollow mesoporous silica spheres (HMS). The FeCu/HMS bimetallic composites were prepared by a direct impregnation and chemical reduction method. The mass ratio of Fe:Cu were 8:0, 6:2, 4:4, 2:6, and 0:8, respectively. And the mass ratio of (Fe + Cu):SiO_2 was 0.08:1. Tests on the catalyst efficiency were carried out at the initial pH of the solutions 5.0, 7.0, and 9.0. Using the $2Fe_6Cu$/HMS catalyst, no differences in the efficiency of dye removal were found (92.6–93%) depending on pH. These differences were shown only in the case when the catalyst was used, in which the share of iron in the mixture was the highest. The optimal parameters in degradation of 100 mg/L orange II is 27.4 mM H_2O_2 with 1.0 g/L of catalyst which was performed at pH of 7.0 and 30°C. It was indicated that by increasing the dose of H_2O_2 from 6.8 to 27.4 mM, the effectiveness of the OGII decomposition process was improved. Further increasing the dose of H_2O_2 reduces the effectiveness due to the scavenging of radicals by excess H_2O_2. Investigating the possibility of reusing the catalyst in subsequent 2-h cycles of OGII decomposition, a slight decrease in efficiency was found. This may be due to the leaching of the elements responsible for its effectiveness from the catalyst or the poisoning of the active sites by intermediate products of the dye decomposition.

The application of the heterogeneous Fenton process to the decomposition of organic compounds is a complex issue, and the process of coming from research work carried out on a laboratory scale to application in real conditions is long and arduous. Scaling technological processes is a real challenge for a wide range of specialists in various fields.

Not without significance is the fact that economic development has resulted in the production of many compounds that can pose a serious threat to the environment. The list of priority substances is very long and is constantly being updated. The problem is not only the prevalence of individual compounds but also the selection of the technology for their removal. The application of the heterocatalytic Fenton process to remove pollutants from solutions of different qualitative and quantitative compositions and to determine the optimal conditions for the process is extremely difficult. Research is being carried out on the decomposition of various compounds, *e.g.*, p-nitrophenol [71], phenolic compounds [72], HCl [73], 1,4-dioxane [74], 4-nitrophenol [75], 2.4-dichlorophenol [76,77], rhodamine B [78–80], and atrazine [81]. Some of them are toxic and can be hazardous to the environment [68].

Organic compounds are difficult to remove and do not biodegrade even if the most effective oxidation processes are being used [82]. However, the study of mixtures of compounds is a much more complicated issue. Industrialists common these compounds in agricultural and industrial products, such as herbicides, fungicides, and wood preservatives [83]. Rodriguez-Gil *et al.* [84] are concerned with using the heterogeneous Fenton process to reduce the pollution of rivers in Madrid, caused by 56 pharmaceutical agents of various chemical groups. Alalm *et al.* [85] showed another example of using the heterogeneous Fenton process, in which they developed a process to reduce the cost of expanding a wastewater treatment plant for a company producing resins. They developed the parameters of the sewage recenter based on parameters such as a daily quantitative database, phenolic tests, and data on the optimal exposure time. The heterogeneous Fenton process has also been used to treat landfill leachate from which natural origin is especially dangerous. It is caused by the fact, that during large-area runoff, rainfall, or intensive decomposition of waste leachate can migrate into the soil and pollute ground or surface water [86–89]. In addition, sludge contains a high concentration of chlorides and toxic heavy metals such as lead, zinc, copper, manganese, and cadmium, which can accumulate in the natural environment [90]. Vilardi *et al.* [91] used the heterogeneous Fenton process to remove pollutants produced in the extractive industry, which contain anionic surfactants such as alkyl benzene sulphonate (ABS) and alkyl ether sulfate (AES). It is used to condition the soil and balance soil pressure [91]. In the HFP, sodium dodecylbenzene sulfonate (SDBS) and linear alkyl benzene sulfonate (LABS) are also being removed [92,93]. Scientists noticed that compound concentration has no significant effect on treatment efficiency [94–96].

Moreover, the Fenton process has been studied as a potential method for the decomposition of microplastics in wastewater treatment plants. Iron is added to the wastewater in the form of iron salts or iron oxide nanoparticles, which act as the catalyst for the Fenton reaction. H_2O_2 is then added to the wastewater to initiate the reaction. The Fenton process has several advantages for the decomposition of microplastics. Firstly, it is an effective method for the degradation of a wide range of organic pollutants, including microplastics. Secondly, it is a relatively simple and cost-effective process that can be easily integrated into existing wastewater treatment plants. Finally, the Fenton process produces no harmful byproducts and is environmentally friendly. However, the Fenton process also has some limitations. Firstly, the reaction can be inhibited by the presence of certain contaminants, such as chloride ions, which can react with the hydroxyl radicals and reduce their effectiveness. Secondly, the Fenton process can also generate large amounts of sludge, which can be difficult to dispose of safely. In summary, the Fenton process with iron catalysts and hydrogen peroxide is a promising method for the decomposition of microplastics in wastewater treatment plants. Further research is needed to optimize the process and address some of its limitations, but it has the potential to be a cost-effective and environmentally friendly solution for microplastic pollution. The Fenton process with iron can be used to decompose different types of microplastics, including polyethylene, polystyrene, and polypropylene. The process works by breaking down the polymer chains in the microplastics into smaller fragments, which can then be further degraded by microorganisms in the wastewater treatment plant. Iron catalysts used in the Fenton process can be in different forms, such as ferrous sulfate, ferric chloride, or iron oxide nanoparticles. The choice of catalyst depends on various factors, including the type and concentration of microplastics, the pH of the wastewater, and the presence of other contaminants. The effectiveness of the Fenton process can be enhanced by optimizing the process parameters, such as the concentration of H_2O_2 and iron catalyst, the reaction time, and the pH of the wastewater. In addition, the use of advanced oxidation processes, such as photo-Fenton or electro-Fenton, can further increase the degradation of microplastics. One of the challenges of using the Fenton process for microplastic decomposition is the need for high concentrations of H_2O_2 and iron catalyst, which can be expensive and require careful handling to ensure safety. In addition, the process can generate heat and release gas bubbles, which can cause foaming and potential clogging in the wastewater treatment system. The Fenton process with iron can also be combined with other treatment processes, such as activated carbon adsorption or membrane filtration, to enhance the overall removal efficiency of microplastics. For example, activated carbon can be used to adsorb the smaller fragments of microplastics that are generated during the Fenton process, while membrane filtration can remove the remaining particles from the treated wastewater [97–99].

The Fenton process with iron is also a potential solution for the removal of microplastics in industrial wastewater, such as those generated from the textile, plastics, or electronics industries. These types of wastewater can contain high concentrations of microplastics and other organic pollutants, which can be difficult to treat using conventional methods.

One of the challenges of using the Fenton process with iron for microplastic decomposition is the potential for the iron catalyst to leach into the environment and cause additional pollution. However, recent studies have shown that iron oxide nanoparticles can be modified to reduce their toxicity and improve their stability in the wastewater treatment process.

The Fenton process with iron works by generating hydroxyl radicals through the reaction between hydrogen peroxide and iron catalysts. These hydroxyl radicals are highly reactive and can break down the polymer chains in microplastics into smaller fragments.

During the Fenton process with iron, the microplastics are exposed to a solution containing hydrogen peroxide and iron catalysts. The hydrogen peroxide reacts with the iron catalysts to produce hydroxyl radicals, which then attack the polymer chains in the microplastics.

The hydroxyl radicals react with the polymer chains in the microplastics through a process called oxidation, which breaks down the chemical bonds and converts the polymers into smaller fragments. These smaller fragments can then be further degraded by microorganisms in the wastewater treatment system or removed through other treatment processes, such as activated carbon adsorption or membrane filtration.

The Fenton process with iron can remove different types of microplastics, including polyethylene, polystyrene, and polypropylene, from wastewater. The efficiency of the process depends on several factors, such as the type and concentration of microplastics, the pH of the wastewater, and the concentration of hydrogen peroxide and iron catalysts.

Overall, the Fenton process with iron catalysts has shown promising results for the decomposition of microplastics in wastewater treatment plants. It is a relatively simple and cost-effective process that can be easily integrated into existing treatment systems. However, further research is needed to optimize the process and address some of the challenges associated with its implementation [98–101].

PROCESS EFFICIENCY IN THE PHARMACEUTICAL, COSMETIC, TEXTILE, AND CHEMISTRY INDUSTRY

The heterogeneous Fenton process is highly effective in target compound decomposition. Figure 5.2 shows that scientists obtained the highest reduction for compounds' concentration in the range of 90–100%.

Figure 5.2 shows a comparison of the efficiency of industries. However, we couldn't be 100% sure about the reliability of research in which scientists obtained a complete degradation of pollutants. As we noticed, it is impossible to break down the compounds to just water and carbon dioxide without using additional processes. Due to the high efficiency of the process and the variability of process conditions, one can't also define the highest compounds removal efficiency. Most of the studies concerned synthetic wastewater, so it is more difficult to remove more types of pollutants and other compounds when it is mixed. Based on the results of the research, we noticed the less treatment degree (73.7–51.2%) for total organic carbon (TOC), but it is also necessary that the value is the mean of different, individual results of research. Figure 5.2 also shows that high efficiency in dye decomposition often occurs together with low efficiency of TOC removal. In Figure 5.2, it is shown the efficiency of degradation TOC in the textile industry is about 51.2%, while in the chemistry industry about 73.7%. In Figure 5.2 we show the efficiency of chemical oxygen demand (COD) removal from 60 to 97.7%. We obtained in our studies the average value of removal COD equal to 71.5%. The results of the method depend on the kind of degradation pollutants and their concentration, time of the process, and process conditions. That's why it is important to check also all parameters of the process, not only their efficiency. Zhang *et al.* [60] noticed the largest difference between the purification of the TOC and compounds. For tartrazine it was about 80% and almost

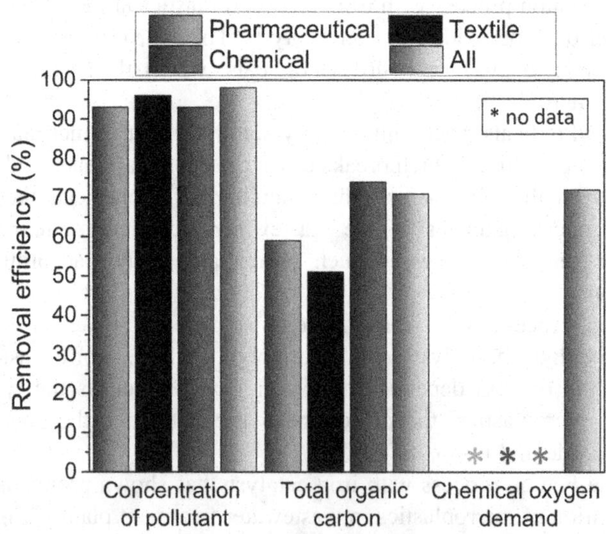

FIGURE 5.2 Efficiency of degradation compounds depends on the analyzed parameter. The figure was created based on rounded results from articles from Table 1.3.

30% of TOC removal. This may be due to the broken bonds/groups/structures responsible for the appearance of color [60]. Gonzalez-Gil *et al.* [84] determined the effectiveness of the heterogeneous Fenton process for water purification in the rivers of Madrid, for which they noted a high (70%) TOC removal of most of the tested compounds from the pharmaceutical industry. However, several compounds with high concentrations of pollutants (4 to 44 mg/L), such as salicylic acid, ofloxacin, caffeine, cotinine, and nicotine were much more difficult to remove from river waters. The scientists obtained the greatest resistance to nicotine at a concentration of 29 to 224 mg/L. On the other hand, the value of COD reduction depended on the studied river (47, 62, and 30%, respectively). These results could indicate oxidizing inorganic compounds in the tested waters, such as chlorides, nitrites, or sulfides, and organic compounds such as chlorine, nitrogen, or sulfur. After heterogeneous Fenton process, most pharmaceuticals tested were below the detection limits and overall reduction of compounds (99.0, 99.1, 98.7, and 99.5%) for the various rivers [84]. In all cases, Gonzales-Gil *et al.* [84] detected some amounts of compounds such as nicotine, caffeine, ofloxacin, cotinine, loratadine, and salicylic acid. Moreover, they also noticed the presence of six other drugs such as hydrochlorothiazide, ciprofloxacin, clarithromycin, erythromycin, citalopram hydrobromide, and carbamazepine [84]. Kakavandi *et al.* [44] studied the degradation of petrochemical wastewater in the heterogeneous Fenton process using magnetite nanoparticles (MNPs@C), obtaining under optimal conditions (1 g/L catalyst and 50 mM H_2O_2 at pH 3.0), the removal of 83.5% COD in 120 min. Most of the compounds present in the wastewater that were determined with GC-MS were almost completely removed. In this case, the differences between the degree of removal of COD and individual compounds may seem small [44]. However, it must be remembered that the GC/MS method can only be used to determine compounds that are at least minimally volatile under the conditions of analysis and resistant to decomposition under the influence of temperature (high temperature in the dispenser should guarantee effective phase change of all substances in the solvent). Thus, it will identify only some of the substances present in the mixture that affect the COD value. In this type of research, it is not possible to accurately determine the degree of mineralization of pollutants in wastewater.

Bogacki *et al.* [12] achieved the highest efficiency of 56.2% TOC removal, after 120 min of treatment for 1:1 H_2O_2/COD mass ratio and 500/500/1,000 mg L^{-1} Fe_3O_4/Fe_2O_3/Fe^0 catalyst doses. In the raw wastewater, 33 substances were determined using the head space-solid phase micro extraction-gas chromatography-mass spectrometry method (HS-SPME-GC-MS). The identified compounds were mainly cosmetic bases (*e.g.*, decamethyltetrasiloxane or decamethylcyclopentasiloxane) and fragrances (*e.g.*, 1,3,4,6,7,8-hexahydro-4,6,6,7,8,8-hexamethyl-cyclopenta(g)-2-benzopyran (galaxolide) or 4-isopropenyl-1-methyl-1-cyclohexene (limonene)). Substances have been detected in raw wastewater that may have a negative impact on the environment, because they have the ability to bioaccumulate and show estrogenic activity, *i.e.*, polycyclic musk, including galaxolide (1,3,4,6,7,8-hexahydro-4, 6,6,7,8, 8,-hexamethyl-cyclopenta[g]benzopyran, HHCB) and tonalide (6-acetyl-1,1,2,4,4,7-hexamethyltetraline,AHTN), UV filters including benzophenone-3 (2-hydroxy-4-methoxyphenyl)-phenylmethanone, BP-3) and 4-MBC (4-methylbenzylidene camphor). These compounds were almost completely removed from the wastewater during treatment using the heterogeneous Fenton process. By obtaining complete removal of substances that are harmful after entering the environment, there is no guarantee that the pre-treated wastewater containing a significant load of pollutants expressed as TOC or COD will be "more safe" for the environment. We do not know what other substances are present in the wastewater, which could not be determined by the HS-SPME-GC-MS method. It is not known what transformations they undergo during the application of the heterogeneous Fenton process. It is also not known whether the decomposition products of substances contained in raw wastewater will not be more harmful than their precursors. In order to obtain such data, it is necessary to supplement research on the effectiveness of the decomposition of industrial wastewater using the heterogeneous Fenton process with ecotoxicity tests or check to what extent wastewater affects biological processes [12]. Muszynski *et al.* [102] studied the effectiveness of combined wastewater treatment processes. The heterogeneous photo-Fenton process using light/Fe^0/H_2O_2 was used to pretreat raw wastewater, thanks to which a 70% COD reduction within 120 min from 1,140 to 341 mg/ L^{-1} was obtained. The chemically pretreated wastewater was biologically treated in a sequencing batch reactor (SBR), with up to 20% volume fraction in the influent, without significant deterioration of COD, nitrogen, and phosphorus removal, but with possible small negative effects on polyphosphate accumulating organisms (PAOs), nitrifiers and other bacteria present in the microbial community. The total effect of wastewater treatment was 97.7% of pollutant removal expressed as COD decrease.

Untreated cosmetic wastewater, subjected to biological treatment in SBR, caused crucial changes in the microbial community structure, leading to a significant decrease in the efficiency of organic carbon, nitrogen, and phosphorus removal. The studies confirm the legitimacy of pre-treatment of industrial wastewater using the heterogeneous Fenton process before discharging it to the sewage system, through which it reaches the municipal sewage treatment plant [102]. In research on the heterogeneous Fenton process, most researchers compare the efficiency of the process determined under optimal conditions. A series of experiments is performed, changing the doses of reagents and the reaction of the initial solution.

If the research concerns processes supported by light of the appropriate wavelength, the so-called dark experiments are also performed for comparison.

Moreover, in Figure 5.3 we have shown, the efficiency of degradation compounds, which depends on the type of process. In a process in which it used only UV light, the efficiency process was the least. Under 20% of the efficiency of the process was noted in operation, which used only hydrogen peroxide. We noted a comparable efficiency in the process, which was carried out with the use of only a catalyst and UV light with hydrogen peroxide. Moreover, we noted the efficiency of dye degradation by 60% in the process, which used hydrogen peroxide and catalysts. The highest efficiency was pointed out in the process, in which additional energy was added as UV, hydrogen peroxide, and catalyst. The course of the process is influenced not only by the type of catalyst but also its dose. In this work, the authors looked at the amounts of catalysts that scientists will most often use for research and recommended for cleaning pollutants, including dyes, which are

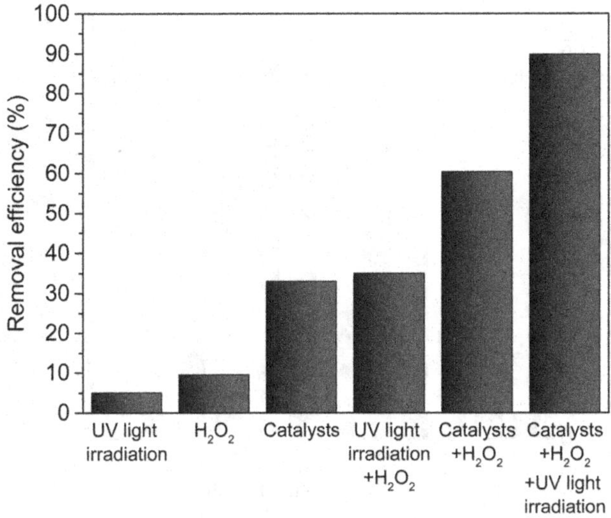

FIGURE 5.3 Efficiency of degradation concentration of compounds depends on the type of process. The figure was created based on rounded results from articles from Table 3.1 (own source).

among the most frequently used standard pollutants, due to their well-known structure and commonly found in industry.

In the previous part, the problem of decomposition of individual dyes was described in detail. However, when research is carried out on real wastewater, the qualitative and quantitative composition of which is complex, it is difficult to determine the optimal conditions. In the wastewater, apart from various types of dyes, there may be other substances that may affect the efficiency of the heterogeneous Fenton process. They can both improve its effectiveness, *e.g.*, through the formation of complexes, or worsen it, as is the case with substances acting as radical scavengers. In studies on the decomposition of substances in real wastewater, the range of H_2O_2 doses is most often determined on the basis of general pollutant indicators such as TOC, COD, and BOD. Sometimes the range of added H_2O_2 doses is determined on the basis of calculations of the theoretical demand for the oxidant based on the knowledge of the process mechanism. Determining the optimal dose of the catalyst poses much greater problems. In practice, higher doses are used in the preliminary research phase. Subsequently, iterations consist in reducing the doses of the catalyst and determining the efficiency of pollutant decomposition. The dose of the catalyst is reduced until the required values of the parameters of the treated wastewater are reached. Bearing in mind that the use of heterogeneous Fenton processes entails the need to incur significant costs, in practice the criterion for selecting the catalyst dose is achieving the required parameters of treated wastewater, the values of which are defined by law. In the case of industrial wastewater containing dyes, the basic criterion is the effective removal of color and increasing the susceptibility of treated wastewater to biological decomposition.

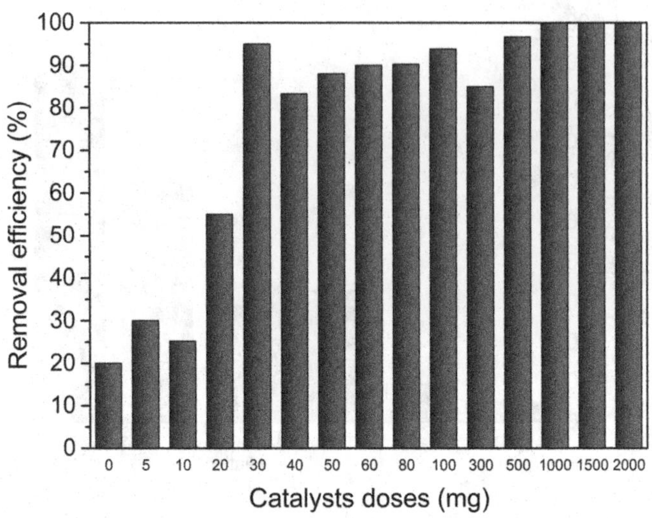

FIGURE 5.4 Dye degradation in heterogeneous Fenton process in relation to catalyst dose. The figure was created based on rounded results from articles from Table 3.1.

Figure 5.4 shows that small doses (5–20 mg/L) are insufficient for dye degradation. Catalyst doses of 30 mg/L are significantly more favorable to use in the Fenton process, which is more effective. When the catalyst dose is bigger than 30 mg/L, the results of treatment reach about 80% dye removal. Figure 5.4 shows that using more than 500 mg/L of catalyst results in a process efficiency higher than 90%, and up to 100% for a 1,000 mg/L catalyst dose.

REFERENCES

1. Zhang, J., et al., Fenton-reaction-stimulative nanoparticles decorated with a reactive-oxygen-species (ROS)-responsive molecular switch for ROS amplification and triple negative breast cancer therapy. *Journal of Materials Chemistry B*, 2019. **7**(45): p. 7141–7151.
2. Ranji-Burachaloo, H., et al., Cancer treatment through nanoparticle-facilitated Fenton reaction. *ACS Nano*, 2018. **12**(12): p. 11819–11837.
3. Shen, Z., et al., Fenton-reaction-acceleratable magnetic nanoparticles for ferroptosis therapy of orthotopic brain tumors. *ACS Nano*, 2018. **12**(11): p. 11355–11365.
4. Filip, J., et al., *Advanced Nano-Bio Technologies for Water and Soil Treatment*. 2020: Springer International Publishing.
5. Sreeja, P.H. and K.J. Sosamony, A comparative study of homogeneous and heterogeneous photo-Fenton process for textile wastewater treatment. *Procedia Technology*, 2016. **24**: p. 217–223.
6. Karthikeyan, S., et al., Treatment of textile wastewater by homogeneous and heterogeneous Fenton oxidation processes. *Desalination*, 2011. **281**: p. 438–445.
7. Punzi, M., B. Mattiasson, and M. Jonstrup, Treatment of synthetic textile wastewater by homogeneous and heterogeneous photo-Fenton oxidation. *Journal of Photochemistry and Photobiology A: Chemistry*, 2012. **248**: p. 30–35.

8. Dantas, T.L.P., et al., Treatment of textile wastewater by heterogeneous Fenton process using a new composite Fe2O3/carbon. *Chemical Engineering Journal*, 2006. **118**(1): p. 77–82.
9. Yu, R.-F., et al., Monitoring of ORP, pH and DO in heterogeneous Fenton oxidation using nZVI as a catalyst for the treatment of azo-dye textile wastewater. *Journal of the Taiwan Institute of Chemical Engineers*, 2014. **45**(3): p. 947–954.
10. Hartmann, M., S. Kullmann, and H. Keller, Wastewater treatment with heterogeneous Fenton-type catalysts based on porous materials. *Journal of Materials Chemistry*, 2010. **20**(41): p. 9002–9017.
11. Sun, L., Y. Li, and A. Li, Treatment of actual chemical wastewater by a heterogeneous Fenton process using natural pyrite. *Int J Environ Res Public Health*, 2015. **12**(11): p. 13762–13778.
12. Bogacki, J., et al., Magnetite, hematite and zero-valent iron as co-catalysts in advanced oxidation processes application for cosmetic wastewater treatment. *Catalysts*, 2021. **11**(1): p. 9.
13. Marcinowski, P., et al., Magnetite and hematite in advanced oxidation processes application for cosmetic wastewater treatment. *Processes*, 2020. **8**(11): p. 1343.
14. Munoz, M., et al., Preparation of magnetite-based catalysts and their application in heterogeneous Fenton oxidation – A review. *Applied Catalysis B: Environmental*, 2015. **176–177**: p. 249–265.
15. Zhou, T., et al., Rapid decolorization and mineralization of simulated textile wastewater in a heterogeneous Fenton like system with/without external energy. *Journal of Hazardous Materials*, 2009. **165**(1): p. 193–199.
16. Hou, B., et al., Three-dimensional heterogeneous electro-Fenton oxidation of biologically pretreated coal gasification wastewater using sludge derived carbon as catalytic particle electrodes and catalyst. *Journal of the Taiwan Institute of Chemical Engineers*, 2016. **60**: p. 352–360.
17. Arslan, İ., I.A. Balcioğlu, and D.W. Bahnemann, Advanced chemical oxidation of reactive dyes in simulated dyehouse effluents by ferrioxalate-Fenton/UV-A and TiO2/UV-A processes. *Dyes and Pigments*, 2000. **47**(3): p. 207–218.
18. Reife, A. and H. Freeman, *Environmental Chemistry of Dyes and Pigments*. 1996: Wiley-Interscience. 352
19. Philips, D., Environmentally friendly, productive and reliable: priorities for cotton dyes and dyeing processes. *Journal of the Society of Dyers and Colourists*, 1996. **112**(7–8): p. 183–186.
20. Wang, J. and J. Tang, Fe-based Fenton-like catalysts for water treatment: Preparation, characterization and modification. *Chemosphere*, 2021. **276**: p. 130177.
21. Clarizia, L., et al., Homogeneous photo-Fenton processes at near neutral pH: A review. *Applied Catalysis B: Environmental*, 2017. **209**: p. 358–371.
22. Chu, J.-H., et al., Application of magnetic biochar derived from food waste in heterogeneous sono-Fenton-like process for removal of organic dyes from aqueous solution. *Journal of Water Process Engineering*, 2020. **37**: p. 101455.
23. Thomas, N., D.D. Dionysiou, and S.C. Pillai, Heterogeneous Fenton catalysts: A review of recent advances. *Journal of Hazardous Materials*, 2021. **404**: p. 124082.
24. Rodrigues, C.S.D., L.M. Madeira, and R.A.R. Boaventura, Optimization and economic analysis of textile wastewater treatment by photo-Fenton process under artificial and simulated solar radiation. *Industrial & Engineering Chemistry Research*, 2013. **52**(37): p. 13313–13324.
25. Hubeny, J., et al., Industrialization as a source of heavy metals and antibiotics which can enhance the antibiotic resistance in wastewater, sewage sludge and river water. *PLOS ONE*, 2021. **16**(6): p. e0252691.

26. Villanueva-Rodríguez, M., et al., Degradation of anti-inflammatory drugs in municipal wastewater by heterogeneous photocatalysis and electro-Fenton process. *Environmental Technology*, 2019. **40**(18): p. 2436–2445.
27. Hussain, S., et al., Enhanced ibuprofen removal by heterogeneous-Fenton process over Cu/ZrO2 and Fe/ZrO2 catalysts. *Journal of Environmental Chemical Engineering*, 2020. **8**(1): p. 103586.
28. Illés, E., et al., Hydroxyl radical induced degradation of ibuprofen. *Science of The Total Environment*, 2013. **447**: p. 286–292.
29. Ellepola, N., et al., A toxicological study on photo-degradation products of environmental ibuprofen: Ecological and human health implications. *Ecotoxicology and Environmental Safety*, 2020. **188**: p. 109892.
30. Adityosulindro, S., C. Julcour, and L. Barthe, Heterogeneous Fenton oxidation using Fe-ZSM5 catalyst for removal of ibuprofen in wastewater. *Journal of Environmental Chemical Engineering*, 2018. **6**(5): p. 5920–5928.
31. Ivanets, A., et al., Heterogeneous Fenton oxidation using magnesium ferrite nanoparticles for ibuprofen removal from wastewater: optimization and kinetics studies. *Journal of Nanomaterials*, 2020. **2020**: p. 8159628.
32. Vorontsov, A.V., Advancing Fenton and photo-Fenton water treatment through the catalyst design. *J Hazard Mater*, 2019. **372**: p. 103–112.
33. Sétifi, N., et al., Heterogeneous Fenton-like oxidation of naproxen using synthesized goethite-montmorillonite nanocomposite. *Journal of Photochemistry and Photobiology A: Chemistry*, 2019. **370**: p. 67–74.
34. Mohammadi, H., B. Bina, and A. Ebrahimi, A novel three-dimensional electro-Fenton system and its application for degradation of anti-inflammatory pharmaceuticals: Modeling and degradation pathways. *Process Safety and Environmental Protection*, 2018. **117**: p. 200–213.
35. Henry Alberto, C.-M., Diclofenac and carbamazepine removal from domestic wastewater using a Constructed Wetland-Solar Photo-Fenton coupled system. *Ecological engineering*, 2020. **v. 153**: p. pp. 105699–2020 v.153.
36. Dinari, M., et al., Construction of new recoverable Ag-Fe3O4@Ca–Al LDH nanohybrids for visible light degradation of piroxicam. *Materials Science and Engineering: B*, 2022. **278**: p. 115630.
37. Bae, S., D. Kim, and W. Lee, Degradation of diclofenac by pyrite catalyzed Fenton oxidation. *Applied Catalysis B: Environmental*, 2013. **134–135**: p. 93–102.
38. Hassani, A., et al., Preparation of magnetite nanoparticles by high-energy planetary ball mill and its application for ciprofloxacin degradation through heterogeneous Fenton process. *J Environ Manage*, 2018. **211**: p. 53–62.
39. Guo, H., et al., Degradation of chloramphenicol by pulsed discharge plasma with heterogeneous Fenton process using Fe3O4 nanocomposites. *Separation and Purification Technology*, 2020. **253**: p. 117540.
40. Luo, T., et al., Efficient degradation of tetracycline by heterogeneous electro-Fenton process using Cu-doped Fe@Fe2O3: Mechanism and degradation pathway. *Chemical Engineering Journal*, 2020. **382**: p. 122970.
41. Benzaquén, T.B., et al., Heterogeneous Fenton reaction for the treatment of ACE in residual waters of pharmacological origin using Fe-SBA-15 nanocomposites. *Molecular Catalysis*, 2020. **481**: p. 110239.
42. Quddus, F., et al., Environmentally benign nanoparticles for the photocatalytic degradation of pharmaceutical drugs. *Catalysts*, 2023. **13**(3): p. 511.
43. Akbari, M.Z., et al., Review of antibiotics treatment by advance oxidation processes. *Environmental Advances*, 2021. **5**: p. 100111.

44. Jaafarzadeh, N., et al., Powder activated carbon/Fe3O4 hybrid composite as a highly efficient heterogeneous catalyst for Fenton oxidation of tetracycline: degradation mechanism and kinetic. *RSC Advances*, 2015. **5**(103): p. 84718–84728.
45. Cao, J., Z. Xiong, and B. Lai, Effect of initial pH on the tetracycline (TC) removal by zero-valent iron: Adsorption, oxidation and reduction. *Chemical Engineering Journal*, 2018. **343**: p. 492–499.
46. Fu, Y., et al., High efficient removal of tetracycline from solution by degradation and flocculation with nanoscale zerovalent iron. *Chemical Engineering Journal*, 2015. **270**: p. 631–640.
47. Zhang, N., et al., Ceria accelerated nanoscale zerovalent iron assisted heterogenous Fenton oxidation of tetracycline. *Chemical Engineering Journal*, 2019. **369**: p. 588–599.
48. Scaria, J. and P.V. Nidheesh, Magnetite–reduced graphene oxide nanocomposite as an efficient heterogeneous Fenton catalyst for the degradation of tetracycline antibiotics. *Environmental Science: Water Research & Technology*, 2022. **8**(6): p. 1261–1276.
49. Zhuang, Y., et al., Enhanced antibiotic removal through a dual-reaction-center Fenton-like process in 3D graphene based hydrogels. *Environmental Science: Nano*, 2019. **6**(2): p. 388–398.
50. Ścieżyńska, D., et al., Application of micron-sized zero-valent iron (ZVI) for decomposition of industrial amaranth dyes. *Materials*, 2023. **16**: p. 1523.
51. Ścieżyńska, D., et al., Waste iron as a robust and ecological catalyst for decomposition industrial dyes under UV irradiation. *Environmental Science and Pollution Research*, 2023. **30**: p. 69024–69041.
52. Ramirez, J.H., et al., Azo-dye Orange II degradation by heterogeneous Fenton-like reaction using carbon-Fe catalysts. *Applied Catalysis B: Environmental*, 2007. **75**(3): p. 312–323.
53. Tu, Y., et al., Heterogeneous photo-Fenton oxidation of Acid Orange II over iron–sewage sludge derived carbon under visible irradiation. *Journal of Chemical Technology & Biotechnology*, 2014. **89**(4): p. 544–551.
54. Lam, F.L.Y., A.C.K. Yip, and X. Hu, Copper/MCM-41 as a highly stable and pH-insensitive heterogeneous photo-Fenton-like catalytic material for the abatement of organic wastewater. *Industrial & Engineering Chemistry Research*, 2007. **46**(10): p. 3328–3333.
55. Ali, M.E.M., T.A. Gad-Allah, and M.I. Badawy, Heterogeneous Fenton process using steel industry wastes for methyl orange degradation. *Applied Water Science*, 2013. **3**(1): p. 263–270.
56. Zhou, L., et al., Fabrication of magnetic carbon composites from peanut shells and its application as a heterogeneous Fenton catalyst in removal of methylene blue. *Applied Surface Science*, 2015. **324**: p. 490–498.
57. Ma, J., et al., Novel magnetic porous carbon spheres derived from chelating resin as a heterogeneous Fenton catalyst for the removal of methylene blue from aqueous solution. *Journal of Colloid and Interface Science*, 2015. **446**: p. 298–306.
58. Yang, S., et al., Degradation of methylene blue by heterogeneous Fenton reaction using titanomagnetite at neutral pH values: process and affecting factors. *Industrial & Engineering Chemistry Research*, 2009. **48**(22): p. 9915–9921.
59. Bel Hadjltaief, H., et al., Influence of operational parameters in the heterogeneous photo-Fenton discoloration of wastewaters in the presence of an iron-pillared Clay. *Industrial & Engineering Chemistry Research*, 2013. **52**(47): p. 16656–16665.
60. Zhang, C., et al., A new type of continuous-flow heterogeneous electro-Fenton reactor for Tartrazine degradation. *Separation and Purification Technology*, 2019. **208**: p. 76–82.

61. Aleksić, M., et al., Heterogeneous Fenton type processes for the degradation of organic dye pollutant in water – the application of zeolite assisted AOPs. *Desalination*, 2010. **257**: p. 22–29.
62. da Fonseca, F., et al., Heterogeneous Fenton process using the mineral hematite for the discolouration of a reactive dye solution. *Brazilian Journal of Chemical Engineering*, 2011. **28**: p. 605–616.
63. Gomes, R.K.M., et al., Treatment of direct black 22 azo dye in led reactor using ferrous sulfate and iron waste for Fenton process: reaction kinetics, toxicity and degradation prediction by artificial neural networks. *Chemical Papers*, 2021. **75**(5): p. 1993–2005.
64. da Silva, A.M., et al., Low-cost flow photoreactor for degradation of Reactive Black 5 dye by UV/H 2 O 2, Fenton and photo-Fenton processes: a performance. *Revista Ambiente & Água*, 2021. **16**(3): p. 1–17.
65. Acisli, O., et al., Combination of ultrasonic and Fenton processes in the presence of magnetite nanostructures prepared by high energy planetary ball mill. *Ultrasonics Sonochemistry*, 2017. **34**: p. 754–762.
66. Saratale, R.G., et al., Hydroxamic acid mediated heterogeneous Fenton-like catalysts for the efficient removal of Acid Red 88, textile wastewater and their phytotoxicity studies. *Ecotoxicol Environ Saf*, 2019. **167**: p. 385–395.
67. Kumar, V., et al., Degradation of mixed dye via heterogeneous Fenton process: Studies of calcination, toxicity evaluation, and kinetics. *Water Environ Res*, 2020. **92**(2): p. 211–221.
68. Duarte, F., et al., Treatment of textile effluents by the heterogeneous Fenton process in a continuous packed-bed reactor using Fe/activated carbon as catalyst. *Chemical Engineering Journal*, 2013. **232**: p. 34–41.
69. Wang, J., et al., Iron–copper bimetallic nanoparticles supported on hollow mesoporous silica spheres: the effect of Fe/Cu ratio on heterogeneous Fenton degradation of a dye. *RSC Advances*, 2016. **6**(59): p. 54623–54635.
70. Wang, Y., et al., Degradation of the azo dye Orange G in a fluidized bed reactor using iron oxide as a heterogeneous photo-Fenton catalyst. *RSC Advances*, 2015. **5**(56): p. 45276–45283.
71. Ban, F.C., X.T. Zheng, and H.Y. Zhang, Photo-assisted heterogeneous Fenton-like process for treatment of PNP wastewater. *Journal of Water, Sanitation and Hygiene for Development*, 2020. **10**(1): p. 136–145.
72. Chakinala, A.G., et al., Industrial wastewater treatment using hydrodynamic cavitation and heterogeneous advanced Fenton processing. *Chemical Engineering Journal*, 2009. **152**(2): p. 498–502.
73. Rahim Pouran, S., et al., Niobium substituted magnetite as a strong heterogeneous Fenton catalyst for wastewater treatment. *Applied Surface Science*, 2015. **351**: p. 175–187.
74. Barndõk, H., et al., Heterogeneous photo-Fenton processes using zero valent iron microspheres for the treatment of wastewaters contaminated with 1,4-dioxane. *Chemical Engineering Journal*, 2016. **284**: p. 112–121.
75. Eshaq, G., et al., Superior performance of FeVO(4)@CeO(2) uniform core-shell nanostructures in heterogeneous Fenton-sonophotocatalytic degradation of 4-nitrophenol. *J Hazard Mater*, 2020. **382**: p. 121059.
76. Zhou, H., et al., Removal of 2,4-dichlorophenol from contaminated soil by a heterogeneous ZVI/EDTA/Air Fenton-like system. *Separation and Purification Technology*, 2014. **132**: p. 346–353.
77. Zhang, C., et al., Heterogeneous electro-Fenton using modified iron–carbon as catalyst for 2,4-dichlorophenol degradation: Influence factors, mechanism and degradation pathway. *Water Research*, 2015. **70**: p. 414–424.

78. Guo, S., et al., Graphene modified iron sludge derived from homogeneous Fenton process as an efficient heterogeneous Fenton catalyst for degradation of organic pollutants. *Microporous and Mesoporous Materials*, 2017. **238**: p. 62–68.
79. Yuan, S.-J. and X.-H. Dai, Facile synthesis of sewage sludge-derived mesoporous material as an efficient and stable heterogeneous catalyst for photo-Fenton reaction. *Applied Catalysis B: Environmental*, 2014. **154–155**: p. 252–258.
80. Zhao, X., et al., Removing organic contaminants with bifunctional iron modified rectorite as efficient adsorbent and visible light photo-Fenton catalyst. *J Hazard Mater*, 2012. **215–216**: p. 57–64.
81. Garrido-Ramírez, E.G., et al., Characterization of nanostructured allophane clays and their use as support of iron species in a heterogeneous electro-Fenton system. *Applied Clay Science*, 2013. **86**: p. 153–161.
82. Malato, S., et al., Decontamination and disinfection of water by solar photocatalysis: Recent overview and trends. *Catalysis Today*, 2009. **147**(1): p. 1–59.
83. Lallai, A. and G. Mura, Biodegradation of 2-chlorophenol in forest soil: effect of inoculation with aerobic sewage sludge. *Environ Toxicol Chem*, 2004. **23**(2): p. 325–330.
84. Rodríguez-Gil, J.L., et al., Heterogeneous photo-Fenton treatment for the reduction of pharmaceutical contamination in Madrid rivers and ecotoxicological evaluation by a miniaturized fern spores bioassay. *Chemosphere*, 2010. **80**(4): p. 381–388.
85. Gar Alalm, M., A. Tawfik, and S. Ookawara, Investigation of optimum conditions and costs estimation for degradation of phenol by solar photo-Fenton process. *Applied Water Science*, 2017. **7**(1): p. 375–382.
86. Aneggi, E., et al., Potential of ceria-based catalysts for the oxidation of landfill leachate by heterogeneous Fenton process. *International Journal of Photoenergy*, 2012. **2012**: p. 694721.
87. Niveditha, S.V. and R. Gandhimathi, Flyash augmented Fe3O4 as a heterogeneous catalyst for degradation of stabilized landfill leachate in Fenton process. *Chemosphere*, 2020. **242**: p. 125189.
88. Sruthi, T., et al., Stabilized landfill leachate treatment using heterogeneous Fenton and electro-Fenton processes. *Chemosphere*, 2018. **210**: p. 38–43.
89. Niveditha, S. and R. Gandhimathi, Mineralization of stabilized landfill leachate by heterogeneous Fenton process with RSM optimization. *Separation Science and Technology*, 2020: p. 1–10.
90. Maiti, S.K., et al., Characterization of leachate and its impact on surface and groundwater quality of a closed dumpsite – a case study at Dhapa, Kolkata, India. *Procedia Environmental Sciences*, 2016. **35**: p. 391–399.
91. Vilardi, G., et al., Heterogeneous nZVI-induced Fenton oxidation process to enhance biodegradability of excavation by-products. *Chemical Engineering Journal*, 2018. **335**: p. 309–320.
92. Takayanagi, A., M. Kobayashi, and Y. Kawase, Removal of anionic surfactant sodium dodecyl benzene sulfonate (SDBS) from wastewaters by zero-valent iron (ZVI): predominant removal mechanism for effective SDBS removal. *Environmental Science and Pollution Research*, 2017. **24**(9): p. 8087–8097.
93. Khorsandi, H., et al., Optimizing linear alkyl benzene sulfonate removal using Fenton oxidation process in Taguchi Method. *Journal of Water Chemistry and Technology*, 2016. **38**(5): p. 266–272.
94. Seibig, S. and R. van Eldik, Kinetics of [Feii(edta)] oxidation by molecular oxygen revisited. New evidence for a multistep mechanism. *Inorganic Chemistry*, 1997. **36**(18): p. 4115–4120.
95. Zang, V. and R. Van Eldik, Kinetics and mechanism of the autoxidation of iron(II) induced through chelation by ethylenediaminetetraacetate and related ligands. *Inorganic Chemistry*, 1990. **29**(9): p. 1705–1711.

96. Cao, M., et al., Remediation of DDTs contaminated soil in a novel Fenton-like system with zero-valent iron. *Chemosphere*, 2013. **90**(8): p. 2303–2308.
97. Piazza, V., et al., Ecosafety screening of photo-Fenton process for the degradation of microplastics in water. *Frontiers in Marine Science*, 2022. **8**.
98. Xing, R., et al., Enhanced degradation of microplastics during sludge composting via microbially-driven Fenton reaction. *Journal of Hazardous Materials*, 2023. **449**: p. 131031.
99. Kim, S., et al., Advanced oxidation processes for microplastics degradation: A recent trend. *Chemical Engineering Journal Advances*, 2022. **9**: p. 100213.
100. Kida, M., S. Ziembowicz, and P. Koszelnik, Impact of a modified Fenton process on the degradation of a component leached from microplastics in bottom sediments. *Catalysts*, 2019. **9**(11): p. 932.
101. Hu, K., et al., Degradation of microplastics by a thermal Fenton reaction. *ACS ES&T Engineering*, 2022. **2**(1): p. 110–120.
102. Muszyński, A., et al., Cosmetic wastewater treatment with combined light/Fe0/H2O2 process coupled with activated sludge. *Journal of Hazardous Materials*, 2019. **378**: p. 120732.

6 Factors Influencing the Effectiveness of the Process

In order to initiate the heterocatalytic Fenton and photo-Fenton processes, it is necessary to provide the initiating factors in the appropriate dose. However, the selection of the dose of reagents and process conditions requires a good knowledge of the mechanism of the Fenton process and photo-Fenton reactions taking place on heterogeneous catalysts. Not without significance is also the knowledge of the decomposition products of compounds present in the reaction mixture [1].

EFFECT OF PH

The pH value is one of the most critical parameters in the heterogeneous Fenton process. The intensity of the reactions, which take place on the catalyst surface is influenced by the environment, especially the pH value in the process [2].

At the very beginning, it is necessary to determine the boundary conditions of pH variation at which the heterogeneous Fenton process can occur. One of the key factors necessary to initiate the process is the presence of Fe^{2+} in the system. The formation of Fe^{2+} is hampered/inhibited due to the formation of $[Fe(H_2O)_6]^{2+}$ and $Fe(OH)_2$, $Fe(OH)_3$ at pH below 3.0 and above 11.0 [2], respectively. Some scientists point to a pH of 2.5 as the lower limit for the effective removal of organic matter in the heterogeneous Fenton process. The effect of pH on the heterogeneous Fenton process depends largely on the type of catalyst itself. Catalysts containing iron atoms in their composition can be divided into two basic groups: unsupported and supported heterogeneous Fenton catalysts. In both types of catalysts, iron may be present as zero-valent iron (ZVI), multimetallic, iron minerals. Clay, carbon, zeolite, polymer, metal organic framework and others can be used as support [3]. The most dependent on initial pH is the heterogeneous Fenton process using the unsupported ZVI catalyst. The processes of decomposition of organic substances using unsupported ZVI in an acidic environment are very effective, because the source of iron ions in the solution is the corrosion process of the ZVI catalyst. Ions are released into the solution and react with H_2O_2, which leads to the production of radicals according to the homogeneous mechanism [3]. At pH greater than 3.0, the process of $HO^\bullet$ radical generation decreases drastically and the process of oxidizing organic substances by Fenton-like oxidation decreases. Conducting the process with the use of ZVI at a higher initial pH of the solution leads to the deposition of corrosion products on the catalyst surface and the formation of passive oxides layer that block the reaction sites on ZVI and hinder the transport of electrons. In this

DOI: 10.1201/9781003364085-6

case, the number of radicals produced and available active sites on the catalyst surface decreases, which results in a decrease in the efficiency of decomposition of organic substances by their oxidation [4]. Efficient oxidation using unsupported ZVI is possible in a narrow range of variation at the initial pH. Obtaining very good pollutant removal effects in a wider pH range can be obtained using unsupported catalysts with iron-containing minerals or bimetallic catalysts. Supported bimetallic/mineral catalyst enable decomposition of organic substances in the widest range of pH. It is possible to run the process at a neutral reaction, thanks to which pH correction is not required after the processes of heterocatalytic oxidation of non-biodegradable organic pollutants before discharging them to the receiver. Wang *et al.* showed that it is possible to effectively decompose orange II at pH 5.0, 7.0, and 9.0, if the share of iron in the supported bimetallic catalyst is lower compared to the content of the second metal [5].

During the decomposition of organic compounds using the heterogeneous Fenton process, products may be formed which, when present in the solution, increase the pH. When using catalysts most sensitive to pH changes, *i.e.*, ZVI, a reaction with dissolved oxygen (DO) takes place and hydroxyl ions are released into the solution. The release of hydroxyl ions is accompanied by the release of iron, which contributes to the formation of further hydroxyl ions [6]. An increase in pH slows down the process of releasing iron from the catalyst.

The pH value also has an important impact on the formation of free radicals, and the size of the surface catalyst, as well as the amount of oxygen, which is dissolved. Another element, in which depend on pH value is the solubility of metals. The pH value parameter also affects the formation of metal complexes with metalloid contamination. During the alkaline process by increasing the adsorption of iron oxides on the surface of the catalyst, metal cations of Cu^{2+} and Cd^{2+} could be removed. Another process allowing metal removal, is the precipitation of metal cations in the form of metal hydroxides. During process operation carried out at pH values below at the point of zero charge (pH_{PZC}), the surface of the catalyst is positive, which can attract negatively charged pollution [7].

The point of zero charge (pH_{PZC}) value is important not only because of the pollutant charge, but also for another reason. For example, pyrite nanoparticles often used in heterogeneous Fenton processes have a pH_{PZC} of 4.6. At a solution pH of 4.6 to 12.0, a positive charge appears on the pyrite surface, which can attract charged molecules including hydroxyl groups (OH^-). The layer of hydroxides hinders the oxidation of pollutants by limiting access to the active sites of the catalyst. Moreover, bearing in mind that in the Fenton-like process, H_2O_2 reacts with Fe^{2+} to form $HO^\bullet$ and OH^-. In addition, the amount of OH^- on the surface of the catalyst significantly reduces the efficiency of the process [8]. The process is highly effective when used in a wide value pH range, while increasing the catalytic efficiency by adding additional external energy. The parameter probe does not significantly affect the direct generation of radicals.

It should also not be forgotten that some pollutants can take different forms depending on the pH of the solution. For example, the decomposition of tetracycline (TC) in the heterogeneous Fenton process using pyrite. TC is an amphoteric compound with one basic functional group (dimethyl amine) and two acidic ones (tricarbonyl amide and phenolic diketine), whose form of presence in solution depends on the reaction. At pH below 3.3 (pK_{a1}), TC exists in the protonated form TCH_3^+. If the pH is below 3.3, it is also below the point of zero charge of pyrite, which has a negative charge on its surface. By increasing the initial pH value of the solution from 3.3 (pK_{a1}) to 4.6 (pH_{PZC}), the efficiency of the process increases, because the charge on the catalyst surface decreases to "0", and the TC molecule has a neutral charge TCH_2^0 at 3.3 (pK_{a1}) < pH < 7.7 (pK_{a2}). At pH 4.6, the highest efficiency of TC decomposition is achieved. Above pH 4.6, the efficiency of the process decreases. A charge appears on the surface of the catalyst and the TC molecule is uncharged up to pH 7.7. Above the pH value of 7.7 (TC pK_{a3}), there is a rapid decrease in catalytic activity, and thus a decrease in TC removal due to the formation of a layer of hydroxides on the catalyst surface and the appearance of a negative charge of the TC molecule (TCH^-). Above pH 9.7 (pK_{a4}), TC is mainly in the form of TC^{2-}) and the degradation process is stopped [8,9]. The situation is even more complicated if the decomposition products include organic substances with weak acidic properties, *i.e.*, derivatives of phenolic compounds, as is the case with the decomposition of TC [8,9]. The resulting products may slightly lower the pH and in some cases increase the efficiency of the process.

Interesting data were presented by Goa *et al.* [7], who used a catalyst with the addition of graphene, obtaining a high degree of purification in a wide range of value pH from 3.03 to 9.44. Zhang *et al.* [10] obtained a high degree of treatment at value pH 6.7 using the modification of the PTTE catalyst. These compounds cause, the high process yields over a broad pH range. Pouran *et al.* [11] noticed that evaluation of adsorption on the catalyst surface with pH value of about 10. Kumar *et al.* [12] noticed a wide pH range, the effectiveness of the process was greater than 90%, including the highest for pH 5.0 (99.82%). As a result, the Fenton process can be used at acidic pH and neutral pH. As the pH value increases, the amount of leached iron decreases. It was noticed the least iron leaching appears at pH 7.0. The high efficiency of the process is due to the acceleration of iron corrosion and the efficient generation of hydroxyl radicals. Efficiency of process also depends on the solubility of the layer of passive oxides on the catalyst surface. At the value pH above 3.0, the concentration of iron complexes increases, the attraction of which between ligands and iron ions in the complexes is high. It hinders the degradation of the compounds and the efficiency of the process low [13,14]. A high value of the pH parameter favors the oxidation of Fe^{2+}. The value pH higher than 4.5 induces Fe^{2+} oxidation followed by rapid precipitation as hydroxide $Fe(OH)_3$. In processes with neutral pH values, precipitation of Fe^{3+} occurs and less efficient production of $^{\bullet}OH$. In the case of the removal of certain dyes, Zhou *et al.* [15] noticed a pH below 3.0 causes the catalyst particles to dissolve rapidly, and slow-down degradation dye. They observed lower levels of oxygen activity at pH below 3.0 due to the combination of hydrogen ions in the Fe^{2+}-EDTA ligand and Fe^{2+}-(EDTA-H) or Fe^{2+}-(EDTA-H_2) ligand. Moreover, the stability of the resulting complexes also depends on the pH. Kwon *et al.* [16] drew an attention to the low efficiency of the process at a pH value of about 2.0, due to the effect of stabilization of H_2O_2. Due to the fact that hydrogen peroxide is a weak acid and is stable at low pH, splitting the hydrogen peroxide into the hydroxyl radical and the hydroxyl ion is difficult in a solution with a pH below 3.0. This process reduces the reactivity H_2O_2 with ferrous ions [16]. Rapid dissolution of metallic iron results in excessive accumulation of hydrogen bubbles at the interface of the catalyst phase. The accumulation of bubbles causes a reduction in the specific surface area of the catalyst, on which it would be possible to remove pollutants. While the value pH of the solution also affects the surface charge, the distribution of iron oxides depends mainly on the pK_a value (constant acid dissociation). The parameter significantly influences the ionization of weak acids or bases. On the other hand, the pH value has a significant influence on the corrosion process of the catalyst [17]. Contrary to the homogeneous Fenton process, which requires the usage of the process at acidic environment, in the heterogeneous process scientists used slightly wider pH range [18,19]. Kumar *et al.* [12] obtained a high degradation result for pH 4.0. The advantage of an acidic pH is that it produces more free radicals. However, more iron leaching occurs during the low pH process [12].

EFFECT H_2O_2 CONCENTRATION

The concentration of the oxidant is considered to be one of the most important factors influencing the effectiveness of the decomposition of organic pollutants with the use of Fe-based catalyst. The concentration of H_2O_2 has a direct impact on the amount of $^{\bullet}OH$ radicals [2]. The dose of peroxide is determined on the basis of the content of the substance that we want to decompose, or the load of pollutants, determined on the basis of indicators such as COD or TOC. Wang *et al.* [20] estimated the amount of H_2O_2 needed to completely decompose 50 mg/L Orange G dye based on the reaction (Eq. 6.1):

$$C_{16}H_{10}N_2Na_2O_7S_2 + 42H_2O_2 \rightarrow 16CO_2 + 2HNO_3 + 2NaHSO_4 + 45H_2O \tag{6.1}$$

According to the reaction, the amount of H_2O_2 needed is 157.8 mg/L. The determined optimal dose was much higher than theoretically calculated [20]. When using too high oxidant dose, the decomposition of the organic substance may be inhibited due to radical scavenging Eqs. (6.2) and (6.3) [5].

$$HO^{\bullet} + H_2O_2 \rightarrow H_2O + HO_2^{\bullet}/O_2^{\bullet} \tag{6.2}$$

$$HO^{\bullet} + HO_2^{\bullet} \rightarrow H_2O + O_2 \tag{6.3}$$

In the case of wastewater, the dose of H_2O_2 is determined on the basis of COD or TOC values. It is assumed that the expected H_2O_2 dose expressed in mg/L corresponds to the COD value expressed in mg O_2/L. This applies to both synthetic and real wastewater. Yu *et al.* [6] studied the degradation of Reactive Black B dye in synthetic wastewater, which was prepared by adding 136.8 mg/L of the dye and 320 mg/L of polyvinyl alcohol to distilled water. The COD of the prepared wastewater was 750–810 mg/L. The use of the optimal dose of H_2O_2 allows to reduce the COD value in treated wastewater by 76% compared to the value determined before the process. The dose needed to effectively remove the color was much lower. This was due to discoloration of the solution, only partial disassembly of the dye [6] was enough.

The main source of radical is hydrogen peroxide, weak acid, characterized by stability in low-value pH. During the pH below 3.0, the cleavage of hydrogen peroxide into the hydroxyl radical and hydroxyl ion is difficult. In high pH value, hydrogen peroxide is not stable and decomposes fast. The required dose of hydrogen peroxide depends on the parameters of the process. Increasing the amount of catalyst, impose a higher dose of hydrogen peroxide [21]. For example, Hadjltaief *et al.* [21] used 0.4 g/L catalysts and 6 mmol/L hydrogen peroxide. Saratale *et al.* [22] used 5 g/L of catalysts and carried out the process at pH of 7.0. They noted a higher value of hydrogen peroxide equal to 40 mM. According to Pouran *et al.* [11] the degradation of methyl blue increased with increasing the dose of H_2O_2. The optimal value of hydrogen peroxide were 800 mg/L [23], 10 mmol/L [7], 10 mmol/L [24], 6 mmol/L, and [25] 5 mmol/L. In the research [26], in which researchers examined wastewater containing basic phenolic compounds, hydrogen peroxide optimal doses were 1,900 mg/L [27], 15.0 mmol/L [28], 105.0 mmol/L, and 1.15 mmol/L [29]. In studies, the value of hydrogen peroxide was 50 mmol/L [30], 8 mmol/L [31], 34.0 mmol/L [32], and 40.0 mmol/L [22].

TEMPERATURE INFLUENCE

The heterogeneous Fenton process does not require any special conditions. Most often, scientists performed research at room temperature and atmospheric pressure. However, the process is still equally effective at a temperature of 10 to 40°C, but the optimal temperature is 30°C. However, the process is also effective in higher temperatures, as it was proven at 40°C [33]. In most processes, a higher temperature is a better parameter for a more efficient reaction. According to Arrhenius kinetics, more incredible energy is vital to the fast and efficient process. However, higher temperatures also have disadvantages. In these conditions, hydroxyl peroxide decomposes intensively to water and oxygen [34]. Araujo *et al.* [23] noticed that an increase in temperature accelerated the decrease in absorbance of the compound. Based on the information obtained in the temperature range of 25–55°C, cation energy of 10.6 kcal/mol was calculated at pH 2.5. [23] Hussain *et al.* [35] stated in their research that the process is more effective when using higher temperatures in the range of 60–80°C. Zhuang *et al.* [34] determined the influence of temperature on the course of the removed COD and efficiency oxidation with ferric oxides to form a catalyst (FeO_x/SBAC). Increasing temperature causes a higher increase in the efficiency of removal COD. However, the process slowed down when temperatures were about 30°C. The efficiency of the process increased 2% when the temperature rose from 30 to 50°C. When the temperature was 70°C, the efficiency of removal of COD dropped by 4.6%. It is important to control all parameters, including temperature. However, controlling temperature can be an additional cost when wastewater treatment is used in industry [34].

REACTION TIME

The reaction rate depends on the concentration of hydrogen peroxide, as well as the surface area of iron oxide on which the process takes place [23]. Adequate time is needed for a successful reaction. The longer we run the process, the greater the efficiency of pollutant breakdown, but only for a limited time. With time, the reagents will start to run out – their concentration decreases, and therefore the speed of the process, hence, the slower loss of COD/TOC. The shorter contact time of the catalyst surface area with hydrogen peroxide the smaller hydroxyl radicals generation and process is less effective [36]. In general, as the process time lengthens, its efficiency increases. The highest efficiency of the process is usually observed for the longest treatment time of 120–180 min. However, it is worth noting that scientists obtained the fastest pollutants removal in the first 30–60 min of the process. It is essential when introducing the process to industry due to energy and water savings. After 60 min of the process, the treatment intensity decreased significantly. We noticed that the TOC values after 60 and 120 min of the process have not changed significantly [37].

Therefore, using the process after 60 min is uneconomical. Longer process time is associated with higher energy consumption or slow down the throughput of the process. The decreasing in the intensity of the process is related to the iron leaching process and its oxidation. Another reason is the lower concentration of hydrogen peroxide, according to its consumption. Without a sufficient dose of hydrogen peroxide, the process does not proceed effectively, which confirms the need to optimize the time of the process, 60 min and it may be the optimal time for the process in the industry. It also noticed that when the process time lengthened, the efficiency grew steadily.

EFFECT OF IONS

The presence of certain ions in the solution, *i.e.*, Cl^-, Br^-, CO_3^{2-}, HCO_3^-, SO_4^{2-}, NO_3^-, $H_2PO_4^-$, reduces the effectiveness of heterogeneous Fenton processes. The reason for the efficiency reduction may be poisoning of the catalyst or reduction of the concentration of reactive oxygen species due to radical scavenging effects [2]. These ions compete with the organics by the $^{\bullet}OH$ radicals, forming other oxidant species ($CO_3^{-\bullet}$, $SO_4^{-\bullet}$, $Cl^{\bullet}$, $Cl_2^{-\bullet}$) according to Eqs. (6.4), (6.5), and (6.6) [38].

$$HCO_3^- + {}^{\bullet}OH \rightarrow CO_3^{-\bullet} + H_2O \tag{6.4}$$

$$CO_3^{2-} + {}^{\bullet}OH \rightarrow CO_3^{-\bullet} + OH^- \tag{6.5}$$

$$SO_4^{2-} + {}^{\bullet}OH \rightarrow SO_4^{-\bullet} + OH^- \tag{6.6}$$

The potential of the resulting radical forms is lower than $^{\bullet}OH$. The effectiveness of the heterogeneous Fenton process in the oxidation of organic pollutants decreases. The degree of reduction depends on the concentration of organic pollutants in real wastewater in relation to the content of ions. The presence of ions in the treated solution requires the use of a higher dose of H_2O_2, of course, to a certain extent (excess H_2O_2 can also act as a $^{\bullet}OH$ radicals' scavenger).

The effect of SO_4^{2-} ions as a radical scavenger $^{\bullet}OH$ is considered weaker than that of HCO_3^- and CO_3^{2-}. This is important when it is necessary to acidify the solution in order to ensure optimal conditions for the decomposition of pollutants in the heterogeneous Fenton process. By using *e.g.*, H_2SO_4 to adjust pH, thus introducing SO_4^{2-} ions, formed as a result of the dissociation of a strong acid. On the other hand, acidification of the solution leads to a disturbance of the carbonate balance and the conversion of carbonates and bicarbonates into CO_2, leading to the removal of radical scavengers from the solution.

Punzi *et al.* [39] studied the degradation of synthetic textile wastewater using the homogeneous and heterogeneous Fenton process. This type of wastewater is characterized by a high content of salts, the total concentration of which ranges from 6 to 40 g/L. Comparing the effectiveness of Remazol Blue RR removal using homogeneous and heterogeneous processes, it was found that a greater negative effect of chloride anions was found in the case of Fe^0. The degree of dye removal was determined on the basis of absorbance changes at the wavelength 290 nm. The presence of NaCl at a concentration of 20 g/L reduced the effectiveness of the process, measured as the degree of absorbance reduction, from 90 to 76% for the homogeneous Fenton process and from 66 to 42% for the heterogeneous Fenton process [39].

EFFECT OF ORGANIC MATER

In the case of solutions with a complex qualitative and quantitative composition, it is very difficult to determine the effect of these substances on the course of the heterogeneous Fenton process. On the one hand, the presence in the composition of these compounds as tert-butyl alcohol or acetylsalicylic acid causes the effect of scavenging radicals. On the other hand, there are chelating substances in wastewater from the cosmetics, food, or textile industry, the presence of which improves the efficiency of treatment using the heterogeneous Fenton process. The most commonly used are ethylenediamine tetra acetic acid (EDTA), ethylenediamine-N,N'-disuccinic acid (EDDS), citric acid (CA), tartaric acid (TA), malonic acid (MA), pyrophosphate (PP), and oxalic acid (OA) [2,40].

In the case of using catalysts from which iron is leached, chelating agents may combine with Fe^{2+} ions and the $HO^{\bullet}$ reaction leads to the formation of Fe^{4+}=O oxoferryl species. The reaction of $HO^{\bullet}$ production may also be accelerated by the formation of complex compounds in which Fe^{2+} ions combine with ligands. The presence of chelating agents is also conducive to the reduction of Fe^{3+} to Fe^{2+} under the influence of light (heterogeneous photo-Fenton process) [40]. When the leaching process does not take place, and the contribution of transformations taking place according to the heterogeneous or semiconductor mechanism of the heterogeneous Fenton process is dominant, the process can be very effective in a wider pH range of the solution. Ligands containing several carboxyl groups at the surface of the catalyst promote the formation of active oxoferryl and Fe^{4+} dihydroxo complexes. As a result, organic compounds decomposition process occurs faster and with greater efficiency. Ligands contain multiple phosphonic acid groups contribute to the stabilization of H_2O_2.

REFERENCES

1. Thomas, N., D.D. Dionysiou, and S.C. Pillai, Heterogeneous Fenton catalysts: A review of recent advances. *Journal of Hazardous Materials*, 2021. **404**: p. 124082.
2. Azfar Shaida, M., et al., Critical analysis of the role of various iron-based heterogeneous catalysts for advanced oxidation processes: A state of the art review. *Journal of Molecular Liquids*, 2023. **374**: p. 121259.
3. Scaria, J., A. Gopinath, and P.V. Nidheesh, A versatile strategy to eliminate emerging contaminants from the aqueous environment: Heterogeneous Fenton process. *Journal of Cleaner Production*, 2021. **278**: p. 124014.
4. Cao, J., Z. Xiong, and B. Lai, Effect of initial pH on the tetracycline (TC) removal by zero-valent iron: Adsorption, oxidation and reduction. *Chemical Engineering Journal*, 2018. **343**: p. 492–499.
5. Wang, J., et al., Iron–copper bimetallic nanoparticles supported on hollow mesoporous silica spheres: the effect of Fe/Cu ratio on heterogeneous Fenton degradation of a dye. *RSC Advances*, 2016. **6**(59): p. 54623–54635.
6. Yu, R.-F., et al., Monitoring of ORP, pH and DO in heterogeneous Fenton oxidation using nZVI as a catalyst for the treatment of azo-dye textile wastewater. *Journal of the Taiwan Institute of Chemical Engineers*, 2014. **45**(3): p. 947–954.
7. Guo, S., et al., Graphene modified iron sludge derived from homogeneous Fenton process as an efficient heterogeneous Fenton catalyst for degradation of organic pollutants. *Microporous and Mesoporous Materials*, 2017. **238**: p. 62–68.
8. Mashayekh-Salehi, A., et al., Use of mine waste for H2O2-assisted heterogeneous Fenton-like degradation of tetracycline by natural pyrite nanoparticles: Catalyst characterization, degradation mechanism, operational parameters and cytotoxicity assessment. *Journal of Cleaner Production*, 2021. **291**: p. 125235.
9. Fu, Y., et al., High efficient removal of tetracycline from solution by degradation and flocculation with nanoscale zerovalent iron. *Chemical Engineering Journal*, 2015. **270**: p. 631–640.
10. Zhang, C., et al., Heterogeneous electro-Fenton using modified iron–carbon as catalyst for 2,4-dichlorophenol degradation: Influence factors, mechanism and degradation pathway. *Water Research*, 2015. **70**: p. 414–424.
11. Rahim Pouran, S., et al., Niobium substituted magnetite as a strong heterogeneous Fenton catalyst for wastewater treatment. *Applied Surface Science*, 2015. **351**: p. 175–187.
12. Kumar, V., et al., Degradation of mixed dye via heterogeneous Fenton process: Studies of calcination, toxicity evaluation, and kinetics. *Water Environ Res*, 2020. **92**(2): p. 211–221.
13. Pignatello, J.J., E. Oliveros, and A. MacKay, Advanced oxidation processes for organic contaminant destruction based on the Fenton reaction and related chemistry. *Critical Reviews in Environmental Science and Technology*, 2006. **36**(1): p. 1–84.
14. Neyens, E. and J. Baeyens, A review of classic Fenton's peroxidation as an advanced oxidation technique. *J Hazard Mater*, 2003. **98**(1–3): p. 33–50.
15. Zhou, T., et al., Rapid decolorization and mineralization of simulated textile wastewater in a heterogeneous Fenton like system with/without external energy. *Journal of Hazardous Materials*, 2009. **165**(1): p. 193–199.
16. Kwon, B.G., et al., Characteristics of p-chlorophenol oxidation by Fenton's reagent. *Water Research*, 1999. **33**(9): p. 2110–2118.
17. Rezaei, F. and D. Vione, Effect of pH on zero valent iron performance in heterogeneous Fenton and Fenton-like processes: a review. *Molecules*, 2018. **23**(12): p. 3127.
18. Zhang, H., H.J. Choi, and C.-P. Huang, Optimization of Fenton process for the treatment of landfill leachate. *Journal of Hazardous Materials*, 2005. **125**(1): p. 166–174.

19. Cheng, M., et al., Visible-light-assisted degradation of dye pollutants over Fe(III)-loaded resin in the presence of H2O2 at neutral pH values. *Environmental Science & Technology*, 2004. **38**(5): p. 1569–1575.
20. Wang, Y., et al., Degradation of the azo dye Orange G in a fluidized bed reactor using iron oxide as a heterogeneous photo-Fenton catalyst. *RSC Advances*, 2015. **5**(56): p. 45276–45283.
21. Bel Hadjltaief, H., et al., Influence of operational parameters in the heterogeneous photo-Fenton discoloration of wastewaters in the presence of an iron-pillared clay. *Industrial & Engineering Chemistry Research*, 2013. **52**(47): p. 16656–16665.
22. Saratale, R.G., et al., Hydroxamic acid mediated heterogeneous Fenton-like catalysts for the efficient removal of Acid Red 88, textile wastewater and their phytotoxicity studies. *Ecotoxicol Environ Saf*, 2019. **167**: p. 385–395.
23. da Fonseca, F., et al., Heterogeneous Fenton process using the mineral hematite for the discolouration of a reactive dye solution. *Brazilian Journal of Chemical Engineering*, 2011. **28**: p. 605–616.
24. Ramirez, J.H., et al., Azo-dye Orange II degradation by heterogeneous Fenton-like reaction using carbon-Fe catalysts. *Applied Catalysis B: Environmental*, 2007. **75**(3): p. 312–323.
25. Ma, J., et al., Novel magnetic porous carbon spheres derived from chelating resin as a heterogeneous Fenton catalyst for the removal of methylene blue from aqueous solution. *Journal of Colloid and Interface Science*, 2015. **446**: p. 298–306.
26. Chakinala, A.G., et al., Industrial wastewater treatment using hydrodynamic cavitation and heterogeneous advanced Fenton processing. *Chemical Engineering Journal*, 2009. **152**(2): p. 498–502.
27. Tu, Y., et al., Heterogeneous photo-Fenton oxidation of Acid Orange II over iron–sewage sludge derived carbon under visible irradiation. *Journal of Chemical Technology & Biotechnology*, 2014. **89**(4): p. 544–551.
28. Liu, F., et al., Application of heterogeneous photo-Fenton process for the mineralization of imidacloprid containing wastewater. *Environmental Technology*, 2020. **41**(5): p. 539–546.
29. Ban, F.C., X.T. Zheng, and H.Y. Zhang, Photo-assisted heterogeneous Fenton-like process for treatment of PNP wastewater. *Journal of Water, Sanitation and Hygiene for Development*, 2020. **10**(1): p. 136–145.
30. Sun, L., Y. Li, and A. Li, Treatment of actual chemical wastewater by a heterogeneous Fenton process using natural pyrite. *Int J Environ Res Public Health*, 2015. **12**(11): p. 13762–13778.
31. Hu, J., et al., In-situ Fe-doped g-C3N4 heterogeneous catalyst via photocatalysis-Fenton reaction with enriched photocatalytic performance for removal of complex wastewater. *Applied Catalysis B: Environmental*, 2019. **245**: p. 130–142.
32. Ali, M.E.M., T.A. Gad-Allah, and M.I. Badawy, Heterogeneous Fenton process using steel industry wastes for methyl orange degradation. Applied Water *Science*, 2013. **3**(1): p. 263–270.
33. Haber, F., J. Weiss, and W.J. Pope, The catalytic decomposition of hydrogen peroxide by iron salts. *Proceedings of the Royal Society of London. Series A - Mathematical and Physical Sciences*, 1934. **147**(861): p. 332–351.
34. Zhuang, H., et al., Advanced treatment of biologically pretreated coal gasification wastewater by a novel heterogeneous Fenton oxidation process. *J Environ Sci (China)*, 2015. **33**: p. 12–20.
35. Hussain, S., et al., Enhanced ibuprofen removal by heterogeneous-Fenton process over Cu/ZrO2 and Fe/ZrO2 catalysts. *Journal of Environmental Chemical Engineering*, 2020. **8**(1): p. 103586.

36. Parolini, M., A. Pedriali, and A. Binelli, Application of a biomarker response index for ranking the toxicity of five pharmaceutical and personal care products (PPCPs) to the bivalve Dreissena polymorpha. *Arch Environ Contam Toxicol*, 2013. **64**(3): p. 439–447.
37. Bogacki, J., et al., Magnetite, hematite and zero-valent iron as co-catalysts in advanced oxidation processes application for cosmetic wastewater treatment. *Catalysts*, 2021. **11**(1): p. 9.
38. Jiménez-Bambague, E.M., et al., Photo-Fenton and Electro-Fenton performance for the removal of pharmaceutical compounds in real urban wastewater. *Electrochimica Acta*, 2023. **442**: p. 141905.
39. Punzi, M., B. Mattiasson, and M. Jonstrup, Treatment of synthetic textile wastewater by homogeneous and heterogeneous photo-Fenton oxidation. *Journal of Photochemistry and Photobiology A: Chemistry*, 2012. **248**: p. 30–35.
40. Vorontsov, A.V., Advancing Fenton and photo-Fenton water treatment through the catalyst design. *J Hazard Mater*, 2019. **372**: p. 103–112.

7 Reactors in the Heterogeneous Fenton Process

One of the advantages of the heterogenenous Fenton process (HFP) is the simplicity of its mechanism. The HFP does not require difficult technologies, devices, or conditions. The process could be carried out in bakers with stirred. Additional elements such as pumps or pH meters are additional devices, that increased the precision and comfort of the carried-out method.

Harney-Ramirez *et al.* [1] prepared one of the most facile forms of the reactor, consisting of a jacketed glass batch reactor, with a capacity of 1.2 L, shown in Figure 7.1. Scientists controlled temperature using a Huber thermostatic bath. The magnetic stirrer was used to mix the solution. Additionally, they used a thermocouple and a pH meter to control the temperature and pH of the solution, respectively. Moreover, the recirculation of the reaction mixture was carried out by a peristaltic pump at a flow rate of ca. 100 mL/min. The results of the HFP were monitored by absorbance measurements, using UV-Vis spectrophotometer with a flow-through cell, and analyzed on a PC [1].

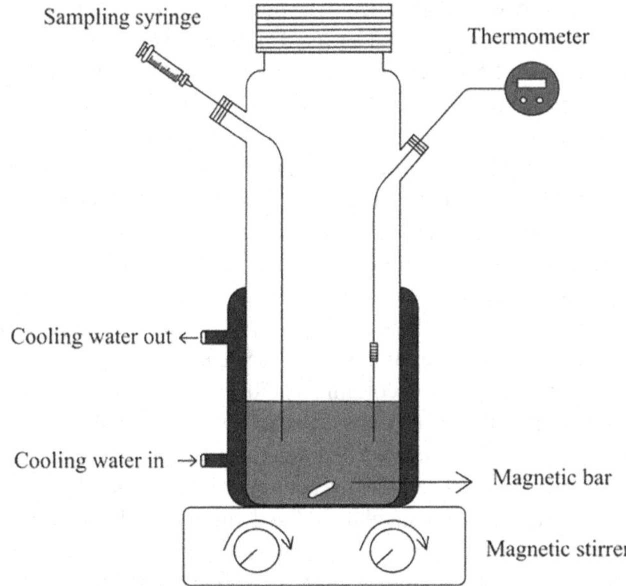

FIGURE 7.1 Schematic of the simplest form of reactor setup.

Source: https://doi.org/10.1021/acs.iecr.1c00976.

DOI: 10.1201/9781003364085-7

However, the reactors have a large variety of designs and modifications due to the type of applied process. Scientists also used additional forms of energy to increase the efficiency and velocity of the heterogeneous Fenton process. Therefore, in our work, we presented the various types of reactors, which are popular and the most comfortable to remove pollutants in laboratories and industries. The types of reactors used in the heterogeneous Fenton process are shown in Table 7.1.

FLUIDIZED-BED REACTOR

One of the most popular forms of the reactor is the fluidized bed (FBR). The process with the FBR is realized on fluidized particles of catalysts, embedded within the reactor, by the fluid flow. To protect particles of catalyst from discharge from the reactor, scientists used the stainless-steel grid, installed on the top of the FBR. This method of decomposing wastewater is characterized by many advantages such as high heat and mass transfer coefficients and well mixing of the solution. Additionally, the recycling and recovery of the catalyst are facile and fast [18]. Moreover, the fluidized beds are characterized by excellent contact between the reagents. Additionally, scientists noticed higher catalyst activity in the FBR than with other methods. The iron particles in the fluidized bed can achieve smooth and steady circulation and improve the efficiency of the process [19].

Aghdasinia *et al.* [18]. carried out the heterogeneous Fenton process in pilot-scale FBR, prepared from polymethyl methacrylate. These reactors are characterized by an internal diameter of 6 cm and 100 cm in height. The schematic of the process is shown in Figure 7.2. To obtain a uniform distribution of the catalyst particles and the fluid, scientists utilized a distributor. The stainless steel distributor of around 0.5 mm was embedded at the bottom of the reactor. To the pumped solution, Aghdasinia *et al.* [18] utilized a 0.3 hp centrifugal pump. The local flow rate was noticed by a rotameter calibrated from 200 to 2,500 L/h for water at room temperature. The process was carried out with 6.6 L of the solution dye and H_2O_2, poured into the storage tank of the reactor. The pH of the solution was adjusted using H_2SO_4 and/or NaOH. Scientists utilized a combination of nano magnetite particle mean sizes comprising 845 mm (80%), 1,300 mm (10%), and 603 mm (10%) in diameter, embedded in the reactor [20].

Duarte *et al.* [3] used a reactor made out of a bed with an iron-based and Fe^{2+} catalyst, deposited on carbon. They packed a borosilicate column (1.5 cm internal diameter and 25.0 cm high), with 6.63 g (10.0 cm) activated carbon impregnated with an iron (AC/Fe) catalyst. Scientists used the inert glass beads at the top and bottom of the column to efficiently disperse the reactants. A peristaltic pump was used to deliver the reagents to the column. The dye concentration in the reactor was monitored continuously for absorbance by a UV-Vis spectrophotometer [3].

Farshchi *et al.* [4] used a fluidized bed reactor, in which the solid particles were fluidized by a liquid or gas. They used reactors with high particle turbulence and a high mass transfer rate between the phases. Farshchi *et al.* [4] also used a cylindrical pump, vessel, and plastic pipes to connect the reactor with plexiglass and a stainless-steel distributor. During the treatment process, a pump-fed the reactor to start the fluidization process. In the next stage, wastewater was pumped into the tank. Vilardi *et al.* [21] used a fluidal bed to sequentially recirculate catalysts. They built the reactor out of a cylindrical plexiglass 100 cm high and an internal diameter of 6.0 cm,

TABLE 7.1
Selected Reactor Used in Heterogenous Fenton Process (Self-prepared)

Type of Reactor	Use Process	Elements of Reactor	Compounds	Reference
Reactor equipped with a plasma-based experimental system	Combination of magnetite-plasma nanocomposites with plasma pulsed discharges (PDP)	Switching power supply, electric detection system, spectrum detection system, spark gap, oscilloscope, voltage probe, needle electrode, reaction chamber, gas purge chamber	CAP	[2]
Fixed-bed reactor filled with a catalyst based on activated carbon impregnated with iron (AC/Fe)	Heterogeneous Fenton process in a continuous packed-bed reactor using Fe/activated carbon as a catalyst	The borosilicate column, packed with AC/Fe catalyst, inert glass spheres at the top and the bottom of the column to promote a homogeneous dispersion of the solution, the column was covered by a water jacket	Alcian Blue-tetrakis (methylpyridinium) chloride	[3]
Fluidized-bed reactor	Natural pyrite catalysts used in fluidized bed reactors for dye degradation in the presence of hydrogen peroxide	Plexiglas, a cylindrical fluidized-bed reactor with stainless steel distributor, pump, inlet, and outlet of the reactor is connected to the pump and reservoir by plastic tubes.	Acid Yellow 36	[4]
Reactor with magnetic sitters	Heterogeneous Fenton process using Fe-SBA-15 nanocomposites	Magnettit sitters, thermostatic bath, liquid sampling device, thermometer, pH control	Acetaminophen (ACE, N-(4-hydroxyphenyl)acetamide)	[5]
Reactor to classical heterogeneous Fenton process	Heterogeneous Fenton-like catalyst biochar modified $CuFeO_2$	The Fenton-like catalytic experiments were carried out in glass conical bottles using a vapor-bathing constant temperature vibrator	Tetracycline	[6]

(Continued)

TABLE 7.1 (Continued)
Selected Reactor Used in Heterogenous Fenton Process (Self-prepared)

Type of Reactor	Use Process	Elements of Reactor	Compounds	Reference
Reactor to electro-Fenton process	Electro-Fenton process with Mineral Iron-Based Natural Catalysts	Undivided and open cylindrical cell, the 3D carbon-felt piece as cathode, and a thin-film BDD on Nb substrate as anode placed at the center of the cell while surrounded by the carbon-felt cathode, magnetic bar to vigorous stirring	Antibiotic CFZ	[7]
Reactor to heterogenous Fenton process with additional energy (ultrasound and ultraviolet)	Rapid decolorization and mineralization of simulated textile wastewater in a heterogeneous Fenton-like system with/without external energy – ultrasounds and ultraviolet	Borosilicate beakers, Pyrex jacket reactor, temperature controller, Sonicator ultrasonic processor, magnetic stirrer, glass diffuser, and fluorescent lamps suspended in a quartz tube located in the center of the photoreactor with aluminum foils	Reactive Black, EDTA	[8]
Reactor to heterogeneous Fenton with ultrasound	Removal of 2,4-dichlorophenol from contaminated soil by a heterogeneous ZVI/EDTA/Air Fenton-like system	Pyrex Jacket Reactor equipped with a two-stage tail gas absorber, magnetic stirrer with a temperature detector, the glass diffuser is regulated by a rotameter	2,4-dichlorophenol	[9]
Reactor to Electro-Fenton	Heterogeneous Electro-Fenton with modified Iron-carbon as the catalyst for 2,4-dichlorophenol Degradation	Undivided cell, ADE was used as the cathode, Ti/IrO_2-RuO_2 as an anode, and modified Fe-C particles as heterogeneous Fenton catalysts	2,4-dichlorophenol	[10]
	Heterogeneous photo-Fenton processes using zero-valent iron	High-pressure mercury immersion lamp from ACE-glass, quartz glass	1,4-dioxane	[11]

Reactor to UV photo-Fenton process and solar photo-Fenton	microspheres for the treatment of wastewater contaminated with 1,4-dioxane	cooling jacket located vertically in the center of the reactor with magnetic stirring, UV-lamp, a solar simulator with a Xe lamp, correction filter, dark screen covering		
Reactor to continuous-flow heterogeneous electro-Fenton	Continuous-flow heterogeneous electro-Fenton (EF) reactor, in which electrochemically formed H_2O_2 was externally injected into the iron-carbon granules packed bed,	The solution tank, fluid flowmeter, pump, the H_2O_2 electrochemical generation system using GDE was used as a cathode, Ti/IrO_2-RuO_2 as an anode, and Na_2SO_4 as the supporting electrolyte cathode, the process reaction cell using a suitable dosage of modified Fe-C particles as catalysts packing at the bottom, the cell was operated in a continuous-flow mode with continuous H_2O_2 inflow from the cell bottom.	Model dye, Tartrazine	[12]
Reactor to cavitation and heterogenous Fenton process	Industrial wastewater treatment using hydrodynamic cavitation and heterogeneous advanced Fenton processing	Feed tank, Controller, plunger pump, price holder, pressure gauge (analog and digital), reactor bed containing iron pieces, immersed in the ice bath, cavitation reactor	Industrial wastewater	[13]
Reactor to three-dimensional heterogeneous electro-Fenton oxidation	Three-dimensional heterogeneous electro-Fenton oxidation using sludge-derived carbon as catalytic particle electrodes and catalyst	Cylindrical plexiglass reactor, one-compartment electrochemical cell, Ti/SnO_2 as the anode, cathode active carbon fiber (ACF), electrolysis cell, reactor to provide oxygen and generate stirring in the solution, thermostatic bath, filter paper	Biologically pretreated coal gasification wastewater	[14]

(Continued)

TABLE 7.1 (Continued)
Selected Reactor Used in Heterogenous Fenton Process (Self-prepared)

Type of Reactor	Use Process	Elements of Reactor	Compounds	Reference
Reactor to heterogenous photo-Fenton process with fluidized	Application of heterogeneous photo-Fenton process for the mineralization of imidacloprid containing wastewater	Three-phase fluidized bed reactor containing under UV irradiation, UV lamp, pyrex tube, glass vessel, (aeration), flowrate, filtred	Imidacloprid containing wastewater	[15]
Reactor to heterogenous Fenton process visible light photo and sunlight	Removing organic contaminants with bifunctional iron-modified rectorite as an efficient adsorbent and visible light photo-Fenton catalyst	Cylindrical Pyrex vessel with a halogen lamp, a cylindrical Pyrex flask, which was surrounded by a circulating water jacket, and a filter placed outside the jacket to remove the wavelength to ensure complete visible light irradiation, the sunshine as the light source	Rhodamine B (RhB)	[16]
Reactor to heterogenous Fenton process with ultrasounds	Insights into the novel application of Fe-MOFs in ultrasound-assisted heterogeneous Fenton system: Efficiency, kinetics, and mechanism	Targeted pollutant to evaluate the US/ Fenton performance of the ultrasonic treatment using an ultrasonic cleaner, filtration with membrane	Tetracycline hydrochloride (TC-HCl)	[17]

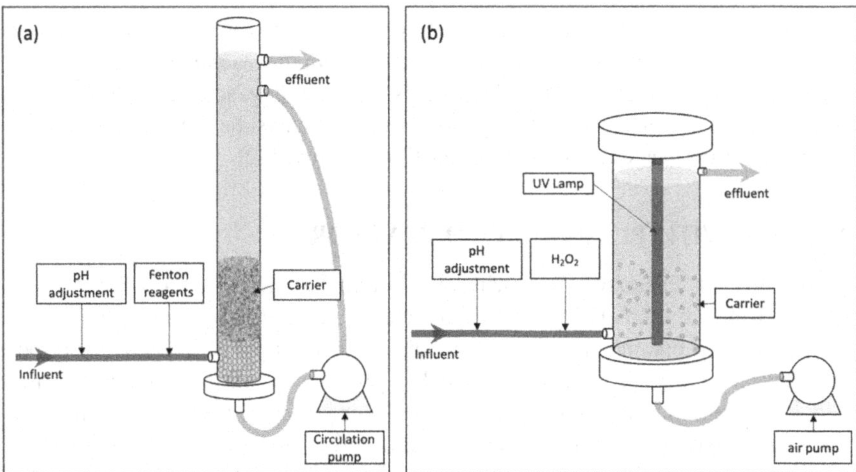

FIGURE 7.2 Schematic set-up of FBR-Fenton system (a) FBR-Fenton-like system with UV photo-Fenton process (b).

Source: https://doi.org/10.1016/j.watres.2020.116692.

a conical dispenser made of perforated stainless steel, and a steel sheet. The reactor was connected to the pump inlet and the outlet to the tank. Additionally, they placed a rotameter to regulate the flow after the pump [21].

Fluidized-Bed Reactor with UV Light

Li *et al.* [19] used a three-phase fluidized bed reactor for N-methyl-2-pyrrolidone mineralization. In their study, the air was sparged continuously from the bottom of the FBR to circulate the Fe^0 throughout the reactor. The process was carried out with a flow rate of 1 L/min. Moreover, scientists utilized the UV lamp, placed vertically in the center of the reactor. Firstly, the air pump was turned on. Secondly, the solution with iron particles, Fe^0 powder, and H_2O_2 was added to the reactor. In the next step, scientists switched on the lamp. Hydrochloric acid and sodium hydroxide were used to maintain the solution's pH. Additionally, the equipment was immersed in a water bath to control the process temperature.

Stream Reactor Apparatus and Catalyst Bed

Dinarvand *et al.* [22] used a round-bottom three-necked flask made of Pyrex glass as a reactor. Scientists carried out the heterogeneous Fenton process utilizing a two-stream impinging reactor with a cylindrical vessel prepared from quartz. Moreover, they used two pneumatic nozzles, placed on the same axis in front of each other. Using a liquid pump, scientists sprained two jets of solutions, with a flow rate of 450 mL/min. Both streams collided in the impingement zone at the catalyst bed. Scientists obtained to hold the catalyst particles at the center of the reactor and had two parallel reticulated wire plates, placed 2–3 mm apart from each other. The solution and oxidizing agent were applied in the batch reactor, while catalysts were loaded in the cage, and placed at the canter of the reaction vessel. The first step of the process was to flow phenol into the feed container from the reaction vessel. In the next step, the solution was recycled to the nozzles. Interestingly, scientists also analyzed the influence of the distance of the nozzles from the catalyst bed, the pump flow rate, and the effect of each collision on the reaction progress [22].

REACTOR WITH PLASMA-BASED REACTOR

Another type of reactor was one with an additional source of energy from plasma reaction. Khataee *et al.* [23] prepared their glow discharge plasma reactor with a Pyrex tube with parameters 40 × 5 cm, shown in Figure 7.3. The two electrodes connected to a DC high-voltage were used to generate the plasma. Firstly, scientists dried 1 g of clinoptilolite powder. In the next step, samples were laid on the Pyrex plate in the positive column place of the tube. Before the process, scientists add N_2 to the tube as the plasma-forming gas. Scientists carried out the process in various conditions. The feed gas was introduced into the reactor at different pressure ranges of 20–30 Pa and the time of generating plasma was between 15 to 60 min [23].

Guo *et al.* [2] used a plasma-based reactor experimental system consisting of a pulse power supply, a reactor system equipped with a needle electrode, and an electric and spectrum detection system. They prepared research with pulsating power triggered by a spark gap. The electrical detection system out of an oscilloscope, probe, and voltage probe, was built. The reactor consists of a needle electrode connected to a reaction chamber and a gas purge chamber [2]. Benzaquén *et al.* [5] used a reactor made of magnetic stirrers to ensure proper mixing conditions, as well as ensure an appropriate catalyst suspension. Also, the reactor had a built-in thermostatic bath connected to a temperature controller, liquid sampling device, thermometer, and pH control [5]. The design of new reactors is associated with the continuous development of treatment methods and the extensive use of the process in various industries.

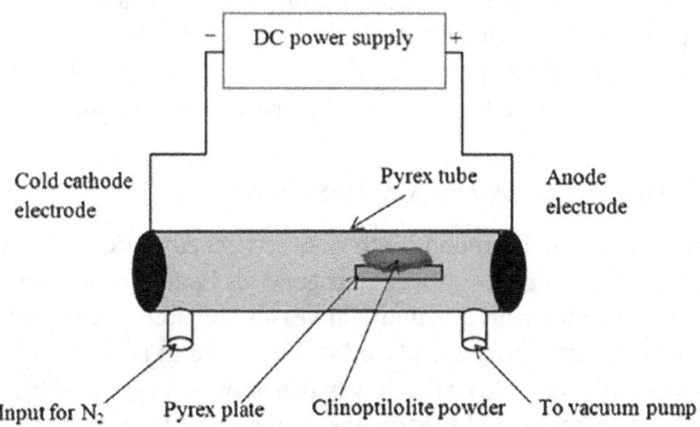

FIGURE 7.3 Reactor with plasma-based reactor.

Source: https://doi.org/10.1021/ie403283n.

ELECTRO-FENTON REACTOR

The electro-Fenton process is characterized by many advantages such as less amount of chemical additives used in the process, easy integration with other processes, and renewable power generation systems. Additionally, scientists noticed that this method is high-yield production of various oxidants such as hydroxyl radical. Moreover, the electro-Fenton could be combined with other systems especially photo-Fenton and anaerobic digestion to increase their efficiency. Therefore, the process could be an excellent method to remove organic pollutants [24].

Behfar *et al.* [25] carried out the photo-electro-Fenton process, with UV radiation, increasing the organic pollutants' mineralization. Moreover, scientists noticed that ultraviolet radiation support radical generation and decompose pollutant difficult to remove through the electro-Fenton process. Behfar *et al.* [25] carried out the process in a 400-mL beaker and two electrodes (anode and cathode) with square shape (2 × 0.5 cm) (Figure 7.4). Both electrodes were made from iron plates with an effective electrode area of 1 cm. Firstly, scientists added wastewater, desired amounts of iron salt (Fe^{2+}), and hydrogen peroxide (H_2O_2) to the reactor. In the next step electrodes were placed in the reactor and solutions were stirred at 400 rpm. Moreover, they utilized the digital DC power supply operated at galvanostatic mode. Additionally, scientists used the UV light lumps with 3, 6, and 9 W to increase the oxidation of the process.

Furthermore, He *et al.* [26] developed an undivided cylindrical glass reactor with a 6 cm diameter and 11 cm height, shown in Figure 7.5. Scientists delivered constant-current electrolyzes with direct current. The cathode used the activated carbon fiber felt (5 × 8 cm), while the anode was a Pt sheet (2 × 2.5 cm). The reactor was located on a magnetic stirrer with a controlled temperature by circulating water. Scientists used the multimeter to measure the current between the cathode and anode. In this process, the nanomagnetic were used as catalysts [26].

Shin *et al.* [24] developed the electro-Fenton process in a one-compartment reactor with magnetic stirring and a volume of 1 L. Scientists utilized a direct current from the power supply (see Figure 7.6). The prepared anodes and cathodes were characterized

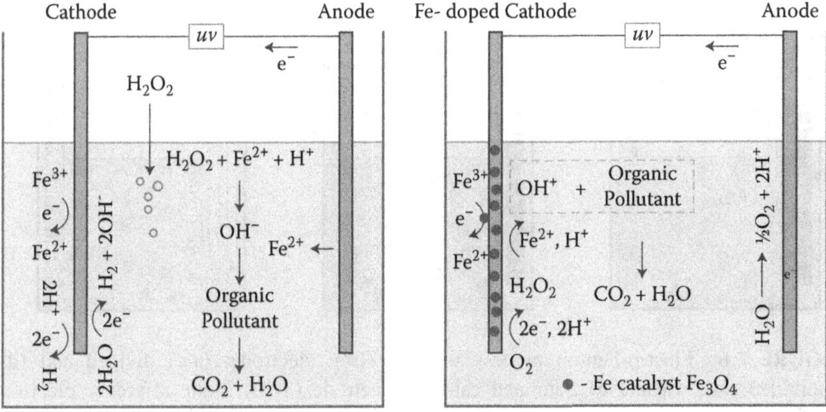

FIGURE 7.4 Electro-Fenton reactor.

Source: https://www.mdpi.com/1996-1944/13/10/2254.

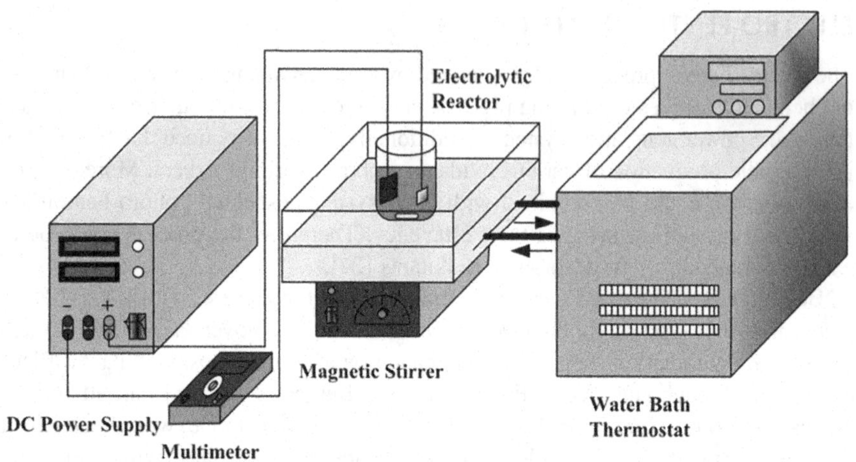

FIGURE 7.5 Electrolytic reactor with water bath thermostat.

Source: https://doi.org/10.1021/ie403947b.

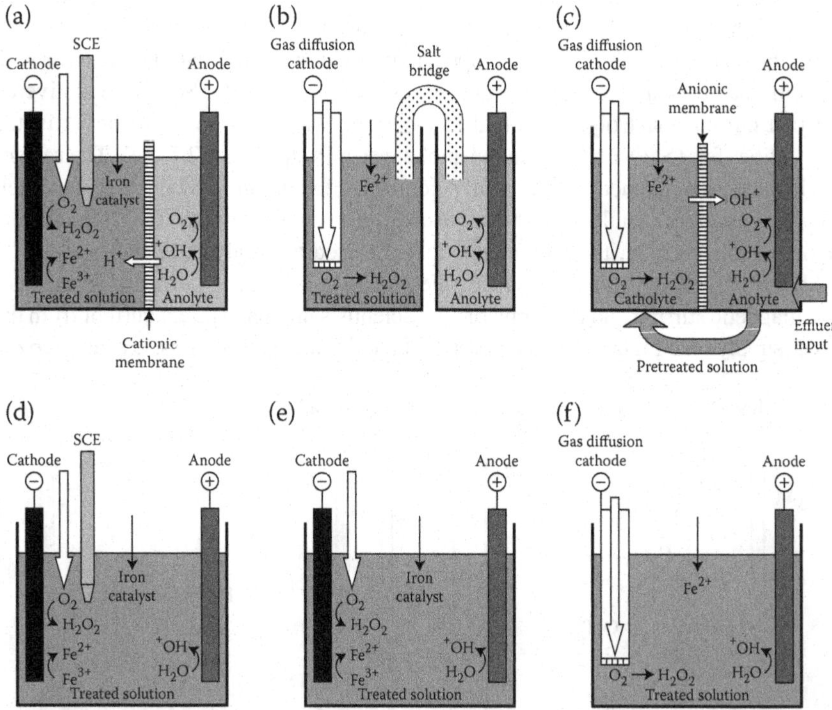

FIGURE 7.6 Electro-Fenton process with the iron electrode: (a-c) divided and (d-f) undivided cells. (a and d) Saturated calomel electrode (SCE) is the reference electrode. (b) Cationic membrane may be used instead of the salt bridge. (c) Solution pretreated in the anolyte is further degraded in the catholyte.

Source: https://doi.org/10.1021/cr900136g.

by a rectangular shape in the form of plates with an active geometric area of 50 cm^2. Both plates were located in parallel with an inter-electrode gap of 1 cm. Scientists used NaCl as a supporting electrolyte. In this process, ferrous ion was transported to the electrolytic reactor for the electrolysis at a sacrificial stainless steel anode. Moreover, the H_2O_2 stock solution was added to the reactor with a constant feeding rate by a peristaltic pump. In this study, the cathode utilized a stainless steel plate of the same size as the anode. Interestingly, the electro-Fenton could be used before electrochemical chlorination [24].

Tang *et al.* [27] carried out electro-Fenton with electrodes prepared of a FeOCl/CC by direct calcination of $FeCl_3{\cdot}6H_2O$. Firstly, scientists immersed the CC in $FeCl_3{\cdot}6H_2O$ ethanol solution and a crucible with ultrasonication to evenly disperse the $FeCl_3{\cdot}6H_2O$. In the final step, the crucible was furnaced at 220°C for 1 h under an air atmosphere. Tang *et al.* [27] used a divided two-compartment reactor of 50-mL volume, carbon plate (2 × 4 cm), and Ag/AgCl for the production of H_2O_2. Moreover, they utilized a 50 mL-beaker, FeOCl/CC (2.5 × 5 cm), carbon plate (2 × 4 cm), and Ag/AgCl to H_2O_2 activation. Additionally, they carried out electro-Fenton with the double-cathode to decompose antibiotics wastewater. In this process, two working electrodes (AGF and FeOCl/CC) were placed between the two compartments in the reactor. In the next step, scientists located the carbon plate counter electrode set 1 cm away from the working electrodes [27].

UV LIGHT AND SOLAR FENTON PROCESS

Scientists still study the adaptation of the Fenton process to industrial. Therefore, they utilized natural, cheap, and ecological energy from UV and visible light. For this purpose, photo-Fenton is one the most interesting method of removing pollutants. However, one of the disadvantages is enough efficiency of the system. In our work, we noticed several types of the Fenton process with UV and sunlight, the efficiency of which in the future can be increased by using a catalyst and the heterogeneous Fenton process.

Farias *et al.* [28] developed an interesting reactor called the insulated solar Fenton hybrid process (see Figure 7.7). To decompose dye, the reactor utilized the

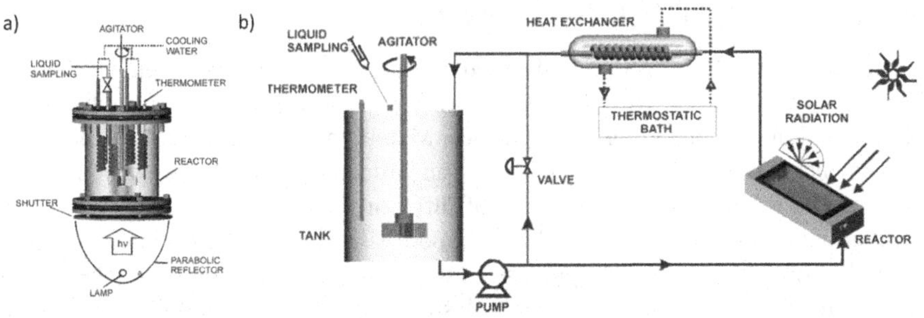

FIGURE 7.7 Schematic representation of the stirred tank laboratory photoreactor.

Source: https://doi.org/10.1021/ie0700258.

entire spectrum of the sun such as solar UV/visible energy in the wavelength region 300–500 nm to activate photochemical reactions and solar thermal energy for wavelengths higher than 500 nm to increase the reaction temperature. Additionally, the reactor was form a new system connecting hybrid photocatalytic-photovoltaic systems. In this method, UV solar radiation is utilized for the photocatalytic decomposition of compounds, while photovoltaic is used to convert visible solar radiation into electricity. In this reactor, the light was delivered from the bottom, using means of a tubular lamp located at the focal axis of a cylindrical reflector. The reactor had also a thermometer, a liquid sampling valve, and a variable-speed stirrer. To facilitate temperature control, scientists connected equipment to the thermostatic bath [28].

Additionally, Farias *et al.* [28] prepared an isothermal, flat-plate solar photoreactor placed inside the loop of a batch recycling system, shown in Figure 7.7. This reactor is characterized by a tank volume higher than the tank of the reactor and a high recirculating flow rate. Moreover, the equipment had excellent mixing in the whole recycling system [28]. Scientists used nylon to prepare the frame and the top of the reactor was made of nylon, while the window was made from Tempax glass. This reactor was completed with a tank of Pyrex glass and a centrifugal pump.

Rodriguez *et al.* [29] prepared the photoreactor with artificial light as a closed cylindrical reactor with a 1 L capacity. The system was equipped with a UV/visible lamp, in which light was the generation from high-pressure mercury vapor, corresponding to an intensity of light of 500 W/m^2. The quartz tube is characterized by special jackets for water recirculation. For temperature control, the system is connected to a thermostatic bath, while a magnetic stir plate was used to mix the solution. Additionally, Rodriguez *et al.* [29] carried out the process in the photoreactor with simulated solar radiation (see Figure 7.8). The photoreactor generated simulated solar radiation. This system is comprised of two reactors. The first of them could close a cylindrical reactor with a jacket for water recirculation from a thermostatic bath for temperature control. The second is characterized by a tubular reactor with a volume of ~0.78 L, located inside the solar box. The solar box was equipped with a xenon lamp of 1,700 W emitted radiation between 300 to 800 nm. The peristaltic pump was used to recirculate wastewater through the two reactors.

MEMBRANES IN THE FENTON PROCESS

Membrane filtration is a commonly used process in wastewater treatment that can be used in combination with the Fenton process with iron for the decomposition of organic pollutants, including microplastics. Membrane filtration involves the use of porous membranes to separate particles and molecules based on their size and charge. The membranes can be made of different materials, such as polymeric or ceramic materials, and can have different pore sizes and configurations.

In the context of the Fenton process with iron, membrane filtration can be used to remove the smaller fragments of microplastics and other organic pollutants that are generated during the Fenton process. These smaller fragments may still be present in the treated wastewater, and can potentially cause environmental harm if released into water bodies.

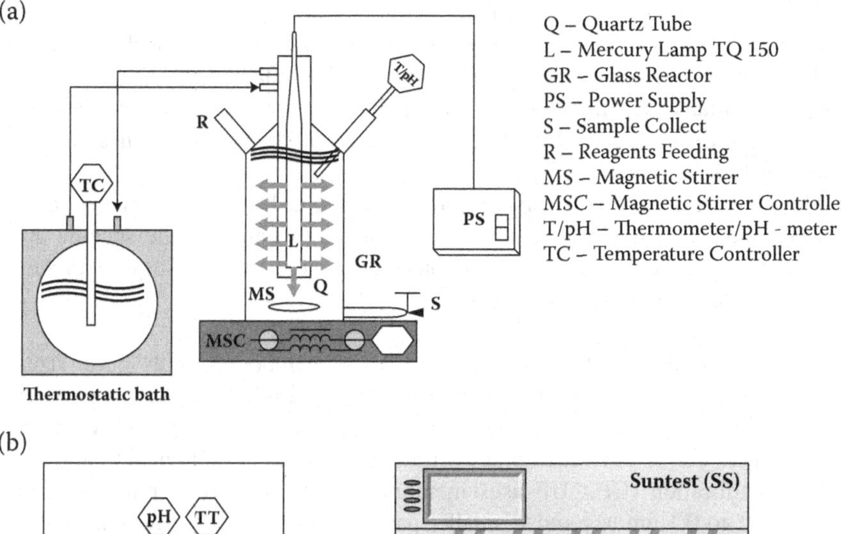

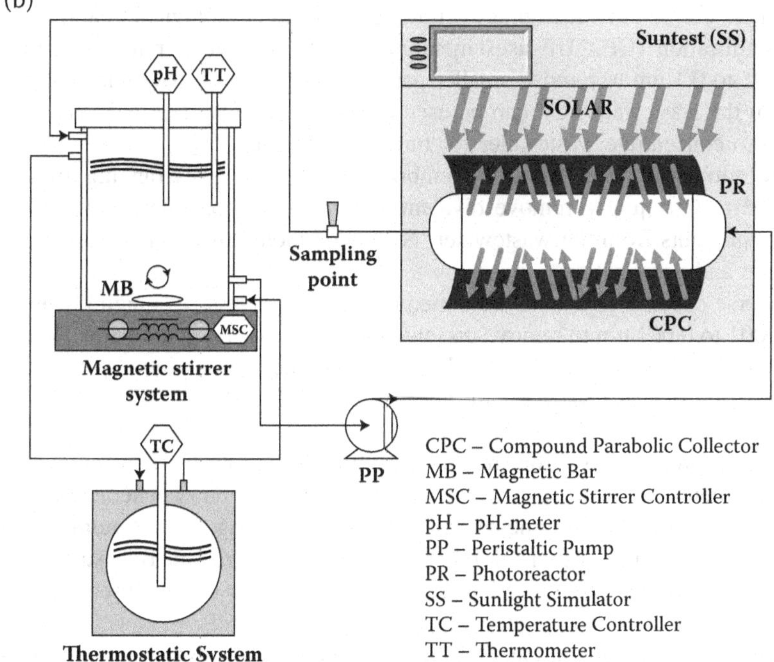

FIGURE 7.8 Diagram of the photoreactor with TO150 Mercury lamp (a) and simulated solar radiation (b).

Source: https://doi.org/10.1021/ie401301h.

By using membrane filtration after the Fenton process with iron, the smaller fragments can be effectively removed from the wastewater. This can result in a more complete removal of microplastics and other organic pollutants and can contribute to a more sustainable and environmentally friendly wastewater treatment system.

Moreover, membrane filtration can also be used in combination with other treatment processes, such as activated carbon adsorption or biological treatment, to further improve the removal efficiency of organic pollutants from wastewater. The use of membrane filtration can also reduce the need for chemical coagulation and sedimentation, which can generate large amounts of sludge and require additional treatment [30–33].

Membrane filtration is a versatile process that can be used in different configurations and with different membrane materials, depending on the specific application and wastewater characteristics. Some common membrane filtration processes used in wastewater treatment include:

- Microfiltration (MF): This process uses membranes with pore sizes ranging from 0.1 to 10 μm to remove suspended solids, bacteria, and other particles from the wastewater. MF is commonly used as a pre-treatment process before other membrane filtration processes or biological treatment.
- Ultrafiltration (UF): UF uses membranes with pore sizes ranging from 0.001 to 0.1 μm to remove smaller particles, viruses, and macromolecules from the wastewater. UF can be used to remove dissolved organic matter and to concentrate wastewater for further treatment.
- Nanofiltration (NF): NF uses membranes with pore sizes ranging from 0.001 to 0.01 μm to remove divalent ions, organic molecules, and other contaminants from the wastewater. NF is commonly used for water softening and desalination.
- Reverse osmosis (RO): RO uses membranes with pore sizes ranging from 0.0001 to 0.001 μm to remove dissolved salts and other contaminants from the wastewater. RO is commonly used for water desalination and concentration of wastewater for further treatment [32–34].

In the Fenton process with iron, membrane filtration can be used to remove the smaller fragments of microplastics and other organic pollutants generated during the process. The smaller fragments can have sizes ranging from nanometers to micrometers, which fall within the range of membrane filtration. By using membranes with appropriate pore sizes, these smaller fragments can be effectively removed from the treated wastewater.

Moreover, the use of membrane filtration in combination with the Fenton process with iron can provide several benefits, such as:

- Increased removal efficiency: The combination of membrane filtration and the Fenton process with iron can result in more complete removal of microplastics and other organic pollutants from the wastewater.
- Reduced chemical consumption: The use of membrane filtration can reduce the need for chemical coagulation and sedimentation, which can generate large amounts of sludge and require additional treatment.
- Improved water quality: The use of membrane filtration can produce high-quality effluent that meets regulatory standards and can be safely discharged into the environment [33,35].

Details on the use of membranes with the Fenton process:

- Membrane fouling: One of the main challenges associated with membrane filtration in wastewater treatment is membrane fouling, which occurs when particles and molecules accumulate on the membrane surface or inside the pores, reducing the membrane's permeability and efficiency. Membrane fouling can be particularly problematic in the Fenton process with iron, as the process generates reactive species that can potentially damage the membrane. To mitigate fouling, pretreatment processes such as coagulation or microfiltration can be employed to remove large particles and protect the membrane surface.
- Membrane materials: The selection of membrane materials is crucial in the Fenton process with iron for wastewater treatment, as different membrane materials have varying chemical and mechanical resistance to the reactive species generated during the process. Membranes made of polymeric materials, such as polyvinylidene fluoride (PVDF), polyethylene (PE), or polypropylene (PP), are commonly used in wastewater treatment applications due to their high permeability and low cost. Ceramic membranes, such as alumina or zirconia, are more resistant to chemical and mechanical stresses and can be used in harsher conditions.
- Membrane configurations: Membrane configurations can also affect the performance and efficiency of the Fenton process with iron in wastewater treatment. The most common membrane configurations used in wastewater treatment include flat-sheet, tubular, and spiral-wound membranes. Each configuration has its advantages and disadvantages in terms of permeability, fouling resistance, and ease of operation.
- Membrane cleaning and maintenance: To ensure the long-term performance and reliability of the membrane filtration system in the Fenton process with iron, proper cleaning and maintenance procedures are essential. Common cleaning methods include backwashing, chemical cleaning, and air scouring, which can help remove accumulated particles and restore the membrane's permeability.
- Membrane integration: Membrane filtration can be integrated with the Fenton process with iron in various ways, depending on the specific treatment goals and wastewater characteristics. For example, membrane bioreactors (MBRs) can combine biological treatment with membrane filtration and the Fenton process with iron for enhanced removal of organic matter and nutrients. Alternatively, the Fenton process with iron can be used as a post-treatment step after membrane filtration to further degrade refractory contaminants and improve water quality.
- Membrane performance monitoring: Continuous monitoring and control of membrane performance is important to maintain optimal treatment efficiency and avoid system failure. Various parameters can be monitored, such as transmembrane pressure, permeate flow rate, and fouling indicators, to detect changes in membrane performance and identify potential problems.

- Membrane fouling control: In addition to pretreatment processes, other strategies can be employed to control membrane fouling in the Fenton process with iron. For example, periodic back pulsing or air sparging can help remove fouling deposits from the membrane surface. Membrane surface modification or coating can also enhance fouling resistance and reduce membrane fouling.
- Membrane selection criteria: The selection of membrane materials and configurations for the Fenton process with iron in wastewater treatment should be based on various criteria, such as chemical compatibility, permeability, fouling resistance, and cost-effectiveness. Other factors, such as membrane pore size, hydrophilicity, and mechanical strength, should also be considered depending on the specific wastewater characteristics and treatment objectives [32–35].

COMBINED FENTON WITH OTHER PROCESSES

The Fenton process could be combined with other processes for wastewater treatment:

- Adsorption: Adsorption is a process that involves the attachment of contaminants onto the surface of a solid material. Combined with the Fenton process with iron, adsorption can enhance the removal of dissolved organic compounds, heavy metals, and other contaminants from wastewater. Various adsorbent materials, such as activated carbon, zeolites, and clay minerals, can be used in combination with the Fenton process with iron to improve treatment efficiency [36].
- Biological treatment: Biological treatment involves the use of microorganisms to degrade or transform contaminants in wastewater. Combined with the Fenton process with iron, biological treatment can enhance the removal of organic matter and nutrients from wastewater, while the Fenton process can further degrade refractory compounds that are resistant to biological treatment. Various biological treatment systems, such as activated sludge, sequencing batch reactors (SBRs), and membrane bioreactors (MBRs), can be used in combination with the Fenton process with iron for wastewater treatment [37].
- Electrochemical treatment: Electrochemical treatment involves the use of an electric field to generate reactive species that can oxidize or reduce contaminants in wastewater. Combined with the Fenton process with iron, electrochemical treatment can enhance the generation of hydroxyl radicals and other reactive species that can further degrade contaminants in wastewater. Various electrochemical treatment technologies, such as electro-Fenton, photoelectro-Fenton, and microbial electrolysis cells (MECs), can be used in combination with the Fenton process with iron for wastewater treatment [38].
- Membrane filtration: Membrane filtration involves the use of a semi-permeable membrane to separate contaminants from water. Combined with the Fenton process with iron, membrane filtration can enhance the removal

of suspended solids and other contaminants from wastewater, while the Fenton process can further degrade dissolved organic compounds that are difficult to remove by membrane filtration. Various membrane filtration technologies, such as microfiltration, ultrafiltration, and nanofiltration, can be used in combination with the Fenton process with iron for wastewater treatment [39].

- Ozonation: Ozonation is a process that involves the use of ozone gas to oxidize contaminants in wastewater. Combined with the Fenton process with iron, ozonation can enhance the generation of hydroxyl radicals and other reactive species that can further degrade contaminants in wastewater. Ozone can also help to remove odors and color from wastewater. Various ozone-based treatment technologies, such as ozone-Fenton, ozone-biological treatment, and ozone-membrane filtration, can be used in combination with the Fenton process with iron for wastewater treatment [40].
- Photocatalysis: Photocatalysis is a process that involves the use of a photocatalyst, such as titanium dioxide (TiO_2), to generate reactive species under UV light that can degrade contaminants in wastewater. Combined with the Fenton process with iron, photocatalysis can enhance the generation of hydroxyl radicals and other reactive species that can further degrade contaminants in wastewater. Various photocatalytic treatment technologies, such as TiO_2-Fenton, TiO_2-biological treatment, and TiO_2-membrane filtration, can be used in combination with the Fenton process with iron for wastewater treatment [37].
- Chemical precipitation: Chemical precipitation is a process that involves the addition of chemicals, such as lime, alum, or ferric chloride, to wastewater to form insoluble precipitates that can be removed by sedimentation or filtration. Combined with the Fenton process with iron, chemical precipitation can enhance the removal of heavy metals and other contaminants from wastewater, while the Fenton process can further degrade organic compounds that are resistant to precipitation. Various chemical precipitation technologies, such as lime-Fenton and ferric chloride-Fenton, can be used in combination with the Fenton process with iron for wastewater treatment [37].

PRE-INDUSTRIAL-SCALE COMBINED SOLAR PHOTO-FENTON

Oller *et al.* [41] developed preindustrial combined solar photo-Fenton with aerobic biological treatment (see Figure 7.9). Scientists purpose to obtain a solution with sufficient biodegradability of the photo-oxidized effluent. Due to removing pollutants, the solution could be poured into an anaerobic immobilized biomass reactor (IBR). Additionally, scientists developed a new hybrid photocatalytic-biological plant with a 4 m^3 daily treatment for real wastewater. The system was built grounds of a pharmaceutical company located in the south of Spain. Oller *et al.* [41] developed the pilot plant of solar photo-Fenton and combined them with the biological treatment of saline non-biodegradable industrial wastewater. The solar photo-Fenton reactor is made for a 3,000 L buffer and the recirculation tank. Moreover, it consisted of a centrifugal

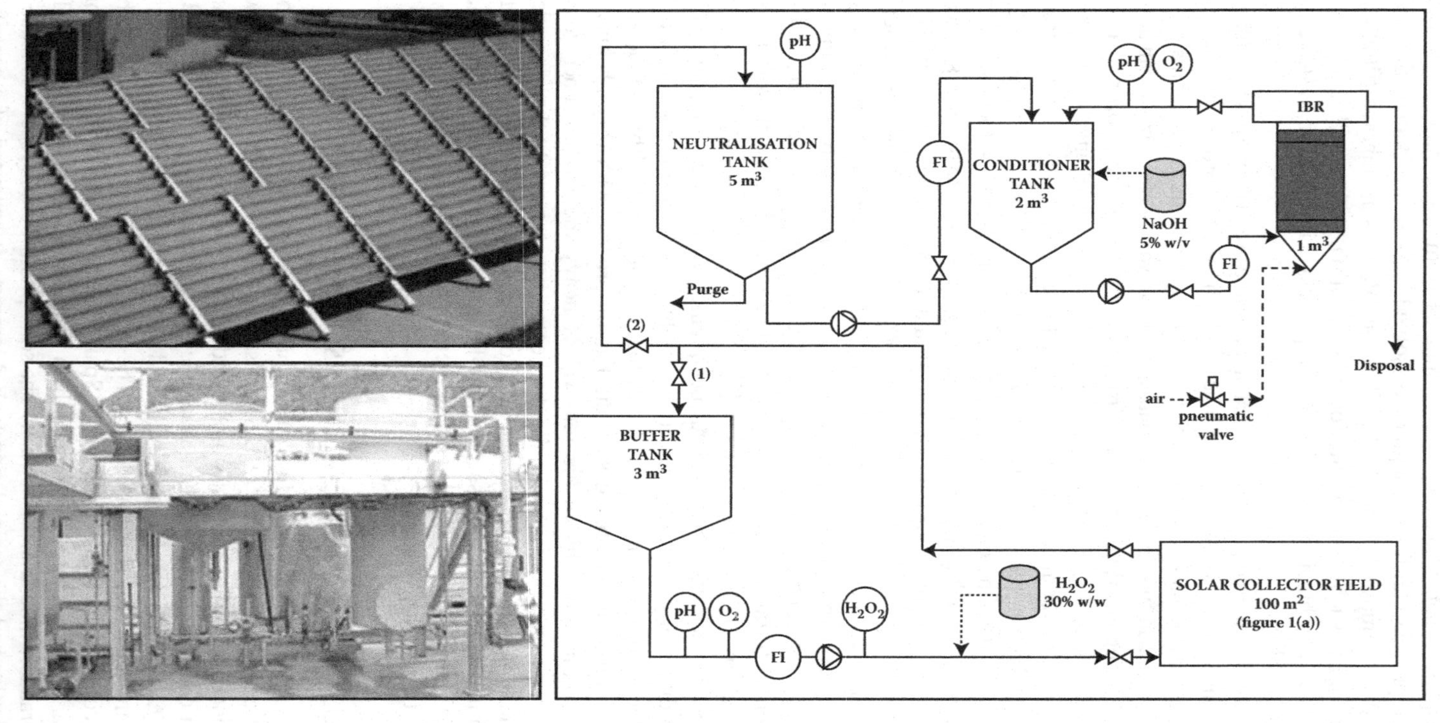

FIGURE 7.9 Views of the combined solar photo-Fenton/biological demonstration plant: solar collector field (left up) and conditioner tank and immobilized biomass reactor (left down). Simplified flow diagram of the demonstration combined photo-Fenton/aerobic biological plant (right).

Source: https://doi.org/10.1021/ie070178v.

pump and a 100 m^2 solar collector field. The collector's system consisted of 15 CPC modules in each row. Additionally, scientists used 50 mm diameter glass absorber tubes located on the frame. The system characterized a total volume of about 4,000 L. Moreover, the sensors of pH, dissolved oxygen, and hydrogen peroxide concentration in the polypropylene were mounted to the equipment. Scientists added to the system the pH and oxygen probe made from PVC-C, while the buffer tank was a conical-bottomed vessel for removing suspended solids. Additionally, the global UV radiometer mounted on a platform was used for the measurement of solar ultraviolet radiation (UV). To obtain and analyze results, the electronic instrument panel in the field and a PC were utilized [41].

REFERENCES

1. Herney-Ramirez, J., et al., Experimental design to optimize the oxidation of Orange II dye solution using a clay-based Fenton-like catalyst. *Industrial & Engineering Chemistry Research*, 2008. **47**(2): p. 284–294.
2. Guo, H., et al., Degradation of chloramphenicol by pulsed discharge plasma with heterogeneous Fenton process using Fe3O4 nanocomposites. *Separation and Purification Technology*, 2020. **253**: p. 117540.
3. Duarte, F., et al., Treatment of textile effluents by the heterogeneous Fenton process in a continuous packed-bed reactor using Fe/activated carbon as catalyst. *Chemical Engineering Journal*, 2013. **232**: p. 34–41.
4. Farshchi, M.E., H. Aghdasinia, and A. Khataee, Modeling of heterogeneous Fenton process for dye degradation in a fluidized-bed reactor: Kinetics and mass transfer. *Journal of Cleaner Production*, 2018. **182**: p. 644–653.
5. Benzaquén, T.B., et al., Heterogeneous Fenton reaction for the treatment of ACE in residual waters of pharmacological origin using Fe-SBA-15 nanocomposites. *Molecular Catalysis*, 2020. **481**: p. 110239.
6. Jain, B., et al., Treatment of pharmaceutical wastewater by heterogeneous Fenton process: an innovative approach. *Nanotechnology for Environmental Engineering*, 2020. **5**(2): p. 13.
7. Heidari, Z., et al., Application of mineral iron-based natural catalysts in electro-Fenton process: a comparative study. *Catalysts*, 2021. **11**(1): p. 57.
8. Zhou, T., et al., Rapid decolorization and mineralization of simulated textile wastewater in a heterogeneous Fenton like system with/without external energy. *Journal of Hazardous Materials*, 2009. **165**(1): p. 193–199.
9. Zhou, H., et al., Removal of 2,4-dichlorophenol from contaminated soil by a heterogeneous ZVI/EDTA/Air Fenton-like system. *Separation and Purification Technology*, 2014. **132**: p. 346–353.
10. Zhang, C., et al., Heterogeneous electro-Fenton using modified iron–carbon as catalyst for 2,4-dichlorophenol degradation: Influence factors, mechanism and degradation pathway. *Water Research*, 2015. **70**: p. 414–424.
11. Barndõk, H., et al., Heterogeneous photo-Fenton processes using zero valent iron microspheres for the treatment of wastewaters contaminated with 1,4-dioxane. *Chemical Engineering Journal*, 2016. **284**: p. 112–121.
12. Zhang, C., et al., A new type of continuous-flow heterogeneous electro-Fenton reactor for Tartrazine degradation. *Separation and Purification Technology*, 2019. **208**: p. 76–82.
13. Chakinala, A.G., et al., Industrial wastewater treatment using hydrodynamic cavitation and heterogeneous advanced Fenton processing. *Chemical Engineering Journal*, 2009. **152**(2): p. 498–502.

14. Hou, B., et al., Three-dimensional heterogeneous electro-Fenton oxidation of biologically pretreated coal gasification wastewater using sludge derived carbon as catalytic particle electrodes and catalyst. *Journal of the Taiwan Institute of Chemical Engineers*, 2016. **60**: p. 352–360.
15. Liu, F., et al., Application of heterogeneous photo-Fenton process for the mineralization of imidacloprid containing wastewater. *Environmental Technology*, 2020. **41**(5): p. 539–546.
16. Zhao, X., et al., Removing organic contaminants with bifunctional iron modified rectorite as efficient adsorbent and visible light photo-Fenton catalyst. *J Hazard Mater*, 2012. **215–216**: p. 57–64.
17. Geng, N., et al., Insights into the novel application of Fe-MOFs in ultrasound-assisted heterogeneous Fenton system: Efficiency, kinetics and mechanism. *Ultrasonics Sonochemistry*, 2021. **72**: p. 105411.
18. Aghdasinia, H., et al., Pilot plant fluidized-bed reactor for degradation of basic blue 3 in heterogeneous fenton process in the presence of natural magnetite. *Environmental Progress & Sustainable Energy*, 2017. **36**(4): p. 1039–1048.
19. Li, H., et al., Mineralization of N-methyl-2-pyrrolidone by UV-assisted advanced Fenton process in a three-phase fluidized bed reactor. *CLEAN – Soil, Air, Water*, 2018. **46**(10): p. 1800307.
20. Aghdasinia, H., et al., Central composite design optimization of pilot plant fluidized-bed heterogeneous Fenton process for degradation of an azo dye. *Environmental Technology*, 2016. **37**(21): p. 2703–2712.
21. Vilardi, G., et al., Heterogeneous nZVI-induced Fenton oxidation process to enhance biodegradability of excavation by-products. *Chemical Engineering Journal*, 2018. **335**: p. 309–320.
22. Dinarvand, M., et al., Degradation of phenol by heterogeneous Fenton process in an impinging streams reactor with catalyst bed. *Asia-Pacific Journal of Chemical Engineering*, 2017. **12**(4): p. 631–639.
23. Khataee, A., et al., Preparation of nanostructured pyrite with N2 glow discharge plasma and study of its catalytic performance in heterogeneous Fenton process. *New J. Chem.*, 2016. **40**.
24. Shin, Y.-U., et al., Sequential combination of electro-Fenton and electrochemical chlorination processes for the treatment of anaerobically-digested food wastewater. *Environmental Science & Technology*, 2017. **51**(18): p. 10700–10710.
25. Behfar, R. and R. Davarnejad, Pharmaceutical wastewater treatment using UV-enhanced electro-Fenton process: Comparative study. *Water Environment Research*, 2019. **91**(11): p. 1526–1536.
26. He, Z., et al., Electro-Fenton process catalyzed by Fe3O4 magnetic nanoparticles for degradation of C.I. reactive Blue 19 in aqueous solution: operating conditions, influence, and mechanism. *Industrial & Engineering Chemistry Research*, 2014. **53**(9): p. 3435–3447.
27. Tang, H., et al., Highly efficient continuous-flow electro-Fenton treatment of antibiotic wastewater using a double-cathode system. *ACS Sustainable Chemistry & Engineering*, 2021. **9**(3): p. 1414–1422.
28. Farias, J., et al., Solar degradation of formic acid: temperature effects on the photo-Fenton reaction. *Industrial & Engineering Chemistry Research*, 2007. **46**(23): p. 7580–7586.
29. Rodriguez, S., A. Santos, and A. Romero, Oxidation of priority and emerging pollutants with persulfate activated by iron: Effect of iron valence and particle size. *Chemical Engineering Journal*, 2017. **318**: p. 197–205.
30. Wang, F., et al., Degradation of the ciprofloxacin antibiotic by photo-Fenton reaction using a Nafion/iron membrane: role of hydroxyl radicals. *Environmental Chemistry Letters*, 2020. **18**(5): p. 1745–1752.

31. Ramírez, J., et al., Heterogeneous photo-electro-Fenton process using different iron supporting materials. *Journal of Applied Electrochemistry*, 2010. **40**(10): p. 1729–1736.
32. Kim, S., et al., Advanced oxidation processes for microplastics degradation: A recent trend. *Chemical Engineering Journal Advances*, 2022. **9**: p. 100213.
33. Farinelli, G., et al., Evaluation of Fenton and modified Fenton oxidation coupled with membrane distillation for produced water treatment: Benefits, challenges, and effluent toxicity. *Science of The Total Environment*, 2021. **796**: p. 148953.
34. Babuponnusami, A. and K. Muthukumar, A review on Fenton and improvements to the Fenton process for wastewater treatment. *Journal of Environmental Chemical Engineering*, 2014. **2**(1): p. 557–572.
35. Sun, M., et al., Nano valent zero iron (NZVI) immobilized CNTs hollow fiber membrane for flow-through heterogeneous Fenton process. *Journal of Environmental Chemical Engineering*, 2022. **10**(3): p. 107806.
36. Lin, R., et al., Synergistic effects of oxidation, coagulation and adsorption in the integrated fenton-based process for wastewater treatment: A review. *Journal of Environmental Management*, 2022. **306**: p. 114460.
37. Ribeiro, J.P. and M.I. Nunes, Recent trends and developments in Fenton processes for industrial wastewater treatment – A critical review. *Environmental Research*, 2021. **197**: p. 110957.
38. Nidheesh, P.V. and R. Gandhimathi, Trends in electro-Fenton process for water and wastewater treatment: An overview. *Desalination*, 2012. **299**: p. 1–15.
39. Rasouli, Y., M. Abbasi, and S.A. Hashemifard, Oily wastewater treatment by adsorption-membrane filtration hybrid process using powdered activated carbon, natural zeolite powder and low cost ceramic membranes. *Water Sci Technol*, 2017. **76**(3–4): p. 895–908.
40. Tanveer, R., et al., Comparison of ozonation, Fenton, and photo-Fenton processes for the treatment of textile dye-bath effluents integrated with electrocoagulation. *Journal of Water Process Engineering*, 2022. **46**: p. 102547.
41. Oller, I., et al., Pre-industrial-scale combined solar photo-Fenton and immobilized biomass activated-sludge biotreatment. *Industrial & Engineering Chemistry Research*, 2007. **46**(23): p. 7467–7475.

8 Kinetics of the Process and Thermodynamical Aspects

Kinetics deals with the study of the rate of reaction, the determination of the influence of various factors on this rate, and, in general, the course of the entire reaction. The study of a given reaction from the kinetic side usually consists in determining the dependence of the rate of formation of products (and consumption of substrates) on the initial concentrations of substrates, temperature, pressure, type of solvent, type, and concentration of the catalyst, etc. The obtained data make it possible to determine the form of the kinetic equation of a chemical reaction and determine the value of its coefficients, which may allow us to understand the reaction mechanism. The laws of chemical thermodynamics allow us to predict only the possible direction of transformations in the system of reactants and the number of individual components present in the system at the moment of reaching equilibrium. With the current state of knowledge, it is possible to describe how the composition of reactants changes over time in a system in which a chemical transformation takes place, mainly by empirical means. Defining the kinetic equation for a given reaction requires experimental data concerning mainly the relationship between the concentration of reactants and reaction rate (see Figure 8.1). The collected experimental data are subjected to theoretical analysis to determine the mechanism, stoichiometry of the reaction, and adjust the appropriate kinetic equation. The rate of a chemical reaction depends on the frequency of the so-called collisions of effective chemical particles, *i.e.*, collisions leading to a reaction.

An effective collision can occur only when the chemical particles are endowed with sufficiently high energy, the so-called activation energy. The rate of a chemical reaction is defined as the increase in the molar concentration of the reaction products or the decrease in the concentration of reactants over time. The instantaneous rate is defined as the rate that defines the slope of the tangent to the curve representing the concentration of the substrate as a function of time. The reaction, which generally consists in changing the number and types of bonds between atoms, requires direct contact between molecules, *i.e.*, the collision of molecules or at least bringing them closer to a distance comparable to the length of bonds. In the case of molecules with a complicated structure, the mutual spatial orientation of colliding particles is also an important component. Often in the kinetic equation, there are only concentrations of some substrates. Sometimes other values like stoichiometric coefficients, powers, and exponents are observed. For complex (multi-stage) reactions, the kinetic equation is a dependence adjusted to the results of experiments and we are usually unable to make molecular sense. Only for the

DOI: 10.1201/9781003364085-8

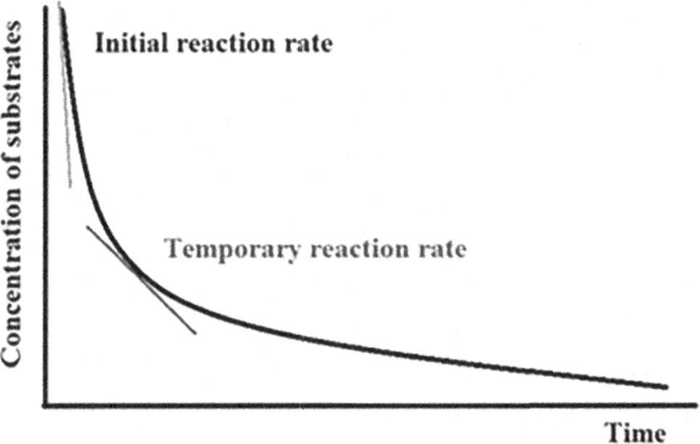

FIGURE 8.1 The concept of changing the rate of a chemical reaction over time.

Source: Own work.

reaction, the form of the kinetic equation is identical to the relationship resulting from the law of mass action. The kinetics are usually described in terms of the reaction order. The reaction order is defined as the sum of the exponents in the kinetic equation of a chemical reaction in the form of a power monomial. The reaction order defined in this way is sometimes referred to as the outer reaction order to distinguish it from the relative order, determined for one substrate and being the value of a single exponent, appearing in the kinetic equation for the concentration of a given substrate. Most often, zero, first, second, and fractional orders of reactions are defined, and mixed models also appear.

A zero-order reaction is a chemical reaction whose rate does not depend on the concentration of the reactants (the kinetic equation is a zero-degree monomial). The reaction rate (r), expressed as the rate of decrease in the substrate concentration (c), is equal to the reaction rate constant (k) under given external conditions (Eq. 8.1):

$$r = -\frac{dc}{dt} = kc^0 = k \tag{8.1}$$

where:

r – reaction rate,
c – reagent concentration,
k – reaction rate constant.

Zero-order reactions are, inter alia: photochemical reactions, in which each of the absorbed photons is used in the reaction (the rate depends on the size of the supplied energy flux, the reaction is "energetically controlled").

A first-order reaction is a reaction in which the kinetic equation (in the form of a power monomial) of the sum of the exponents is equal to 1. It can also be defined as an elementary reaction whose rate is proportional to the concentration of only one reactant (Eqs. 8.2–8.3).

$$r = kc \tag{8.2}$$

$$\frac{dc}{dt} = -kc \tag{8.3}$$

where:

r–reaction rate,
c – reagent concentration,
k – reaction rate constant.

Characteristic examples of first-order kinetics reactions are single-molecular reactions that are not the result of collisions of molecules of different reactants, like radioactive or thermal decay.

A second-order reaction is a reaction in which the kinetic equation (in power monomial form) the sum of the exponents is equal to 2. It could be described according to Eqs. (8.4–8.5).

$$r = kc^2 \tag{8.4}$$

$$\frac{dc}{dt} = -kc^2 \tag{8.5}$$

where:

r – reaction rate,
c – reagent concentration,
k – reaction rate constant.

Second-order reactions are very common. Examples include, among others: reactions of the formation of hydrogen iodide or the decomposition of nitrogen dioxide.

It may happen that the concentration of one of the reactants remains constant during the reaction (it may be, for example, a catalyst in the process). Its concentration can be entered into the description of the reaction rate constant, leading to a pseudo-first-order (Eq. 8.6) or pseudo-second-order (Eq. 8.7) reaction rate.

$$k' = kcat \tag{8.6}$$

where:

cat – constant catalyst concentration,
k – reaction rate constant,
k′ pseudo-first-order rate constant.

$$k' = kcat^2 \quad (8.7)$$

where:

cat – constant catalyst concentration,
k – reaction rate constant,
k′ pseudo-second-order rate constant.

Many chemical reactions have a complicated mechanism of action. Chain reactions can be mentioned here. For many of them, the exponent is not an integer, but a fraction (0.5, 1.5, etc., Eqs. 8.8–8.9).

$$r = kc^n \quad (8.8)$$

$$\frac{dc}{dt} = -kc^n \quad (8.9)$$

where:

r–reaction rate,
c – reagent concentration,
k – reaction rate constant,
n – reaction order, eg: 0.5, 1.5, etc.

Classical kinetic models are dominated by the assumption that a chemical reaction proceeds to the very end, *i.e.*, until one of the reactants is entirely consumed. In many cases, this assumption is correct, especially in the case of one-, two- or three-component systems. However, chemical reactions very often take place in complex multi-element matrices, whose components can react with each other in a complicated way, causing, among others, inhibition of other reactions taking place in the system. This allows us to assume that only some of the available reagents can undergo a chemical reaction. The residual will remain after the process in the reaction system. This type of phenomenon can be observed particularly often during wastewater treatment processes. Because of that modified first-order (Eq. 8.10) and modified second-order (Eq. 8.11) kinetic models were developed.

$$\frac{dc}{dt} + b = -kc \quad (8.10)$$

where:

r–reaction rate,
c – reagent concentration,
k – reaction rate constant,
b – correction constant, assuming the residual concentration of the reagent.

$$\frac{dc}{dt} + b = -kc^2 \tag{8.11}$$

where:

r–reaction rate,
c – reagent concentration,
k – reaction rate constant,
b – correction constant, assuming the residual concentration of the reagent.

Wastewater is a complex, multi-element mixture; hence, it is extremely difficult to determine the concentration of individual compounds and describe the kinetics of the decomposition reaction for each of them separately. For simplicity, therefore, collective parameters are used, determining the content of compounds with similar characteristics or properties. Among them, BOD, COD, or TOC can be mentioned, in relation to which the decrease in time is most often calculated in the kinetics of the process.

The wide scope of knowledge about the heterogeneous Fenton favors understanding this process and its kinetics, which many researchers constantly develop. The kinetics of the process depends on pH value, oxygen partial pressure, and the initial application of Fe^{2+}. Pouran *et al.* [1] observed that the removal of pollutants is proportional directly to the concentration of the compound, the catalyst in the solution, and the amount of $^{\bullet}OH$ radicals. The rate of decolorization should follow the pseudo-first-order kinetic mechanism due to the constant amount of catalyst and $^{\bullet}OH$ radicals. Pouran *et al.* [1] noticed that the kinetic rate constants decreased with an increasing initial concentration due to the action of the active ingredient on the catalyst surface. The degradation rate of the soil was higher with the greater ratio of active ingredients to dye molecules at low soil concentrations [1]. Zhang *et al.* [2], confirmed that 2,4-dichlorophenol degradation followed pseudo-first-order kinetics ($R_2 > 0.95$). They obtained the value of the constant (K_{obs}) from the slope of the straight line appearing in the first-order plot [2]. Pouran *et al.* [1] determined the kinetic parameters using the linear regression method. They obtained the K_{app} (the apparatus affinity constant under the condition) value from the slope of the line by plotting $-\ln(C_t/C_0)$ against the reaction time. This method is most often used to determine reaction kinetics. The initial value of Fe^{2+} is calculated according to the formula in Eq. (8.12) [3]:

$$d\,[Fe^{2+}](dt)^{-1} = k\ [OH^{-}]^2 PO_2\,[Fe^{2+}] \tag{8.12}$$

Ma *et al.* [4] based on the plot of $\ln(C_0/C)$ as a function of time, also confirmed that the degradation reaction of methylene blue followed pseudo-first-order kinetics. The reaction rate constant was calculated to be 0.1225/min ($R_2 = 0.9906$) [4,5].

Seretele *et al.* [6] described pseudo-first and second-order kinetics using the formulas in Eq. (8.13):

$$ln(C_t - C_0) = -k_1 t \tag{8.13}$$

And pseudo-second-order kinetic equations such as in Eq. (8.14):

$$\frac{1}{C_t} - \frac{1}{C_0} = k_2 t \tag{8.14}$$

C_0 and C_t are dye concentrations (μM) at initial and time t, respectively. Moreover, k_1 (1/min) and k_2 (1/Mmin) are the rate constants of the pseudo-first-order and pseudo-second-order kinetics, respectively [6]. Kumar *et al.* used the kinetics to calculate zero Eq. (8.15) order, first order Eq. (8.16), pseudo-first-order Eq. (8.17), second-order Eq. (8.18), and pseudo-second-order (8.19) [7]:

$$|A_t| - |A_0| = -kt \tag{8.15}$$

$$ln\frac{A_t}{A_0} = -k_1 t \tag{8.16}$$

$$\frac{1}{A_t} - \frac{1}{A_0} = -k_2 t \tag{8.17}$$

$$\log(A_t - A_e) = \log(A_e - A_t)\left(\frac{-k_1}{2.303}\right) \tag{8.18}$$

$$\frac{1}{A_t} - \frac{1}{A_0} = \frac{1}{k_2 A_0^2} \tag{8.19}$$

In the heterogeneous Fenton process, other kinetic models can also be used, including the description of pollutant distribution, such as the Langmuir-Hinshelwood model. This model describes the kinetics of pollutant degradation, which relates to the rate of pollutant degradation (r) and the concentration of the reagent (C) at time t, such as in Eq. (8.20) [8,9].

$$\mathrm{r} = \frac{dc}{dt} = \frac{K_r K_{ad}}{1 + K_{ad} c} \tag{8.20}$$

In which K_r is the rate constant, and K_{ad} is the adsorption equilibrium constant. Wang *et al.* [10] determined the maximum efficiency of the catalyst during the degradation of the pollutant (N-nitrosodimethylamine, NDMA) using the Eq. (8.21):

$$\text{Maximum molar yield}(\%) = \frac{[NDMA]_m}{[M]_0} {}_* 100\% \tag{8.21}$$

where $[NDMA]_m$ (mM) is the maximum catalyst concentration formed during pollution degradation and $[M]_0$ (mM) is the initial catalyst concentration [10].

However, the best fit for experimental values showed a linear Behnajady – Modirshahla – Ghanbery (BMG) method. Regression values the factor was close to 1, which makes it clear that this model best describes dye degradation. In the BMG model, they calculated the value of kinetics and regression according to Eq. (8.22) [7].

$$\frac{t}{1-\frac{A_t}{A_o}} = m + bt \tag{8.22}$$

where m and b are constants of the BMG model relating to oxidation capacity and reaction kinetics. This value was obtained by intercept (m) and slope (b) by plotting a graph between $1 - \frac{A_t}{A_o}$ versus t [7].

It is worth pointing out, that Nakagawa *et al.* [11] performed the complete kinetic modeling of the heterogeneous Fenton process.

We [12] used first- and second-order kinetics Eqs. (8.23–8.24). The reaction kinetics was based on specific substrates and products, but it is hard to analyze a complex group of compounds, *e.g.*, industrial wastewater. Therefore, we [12,13] used collective parameters, such as TOC, to describe the kinetics of the reaction of wastewater treatment. According to the first-order and second-order kinetics, the reaction continues until the substrate removes such as TOC. We [12,13] introduced modifications to first- and second-order kinetics formulas. However, some compounds are difficult to remove. The alterations of reaction rely on adding value *b* to Eqs. (8.25–8.26), which means the amount of unremoved TOC. The accuracy of the equations showed that the most effective was to keep the first-order equation with modification [12,13].

$$TOC = (TOC_0 \times e^{-kt}) \tag{8.23}$$

$$\mathrm{TOC} = (\mathrm{k} \times \mathrm{t} + 1((\mathrm{TOC}_0)^{-1})^{-1} \tag{8.24}$$

$$TOC = (TOC_0 - b) \times e^{-kt} + b \tag{8.25}$$

$$TOC = (kt + (TOC_0 - b)^{-1})^{-1} + b \tag{8.26}$$

Saratale *et al.* [6] used the Arrhenius formula to calculate the activation energy of dye removal in their research (8.27):

$$lnK_2 = -\frac{E_a}{RT} + lnA \tag{8.27}$$

E_a, R, T, and A refer to the activation energy, universal gas constant absolute temperature, and pre-exponential factor, respectively. Saratale *et al.* [6] used the linear form of the equation to calculate the enthalpy and entropy of activation in their research, such as in Eq. (8.28):

$$ln\frac{K_2}{t} = ln\frac{K_B}{h} + \frac{\Delta S^{\neq}}{R} - \frac{\Delta H^{\neq}}{RT} \tag{8.28}$$

where K_b, h, $\Delta S^{\neq}$, and $\Delta H^{\neq}$ are Boltzmann constant, Planck constant, activation entropy, and activation enthalpy, respectively. Saratale *et al.* [6] used linear regressions to calculate Ea, ΔH, and ΔS, obtaining values of 25.29 kJ/mol, 22.64 kJ/mol′ and −188.69 J/molK. The obtained results (ΔH value) indicate that mentioned reaction is endothermic [14]. Chen *et al.* [15] 31 kJ/mol and Lin *et al.* [16] obtained lower activation energy in their research, obtaining 32.8 kJ/mol. The initial energy of the reaction depends significantly on the catalyst used. In the research of Saratele *et al.* [6], with higher initial reaction energy, less energy is needed to activate the HpOFe-GAC catalyst (iron-H_{po} ligand catalyst supported on granular activated carbon) [6].

Kumar *et al.* [7], in his research, calculated Gibbs free energy (ΔG) (8.29), activation energy (E_a) (8.29), activation enthalpy (ΔH°), and activation entropy (ΔS°) (8.31).

$$\Delta G = -RTlnk \tag{8.29}$$

$$lnk = lnk_0 - \frac{E_a}{RT} \tag{8.30}$$

$$ln\frac{k}{t} = ln\frac{R}{N_a h} + \frac{\Delta S^{\circ}}{R} - \frac{\Delta H^{\circ}}{RT} \tag{8.31}$$

where lnk_0 is Arrhenius constant, k is the kinetic rate constant, R is the gas constant, E_a is aviation energy, N_a is Avogadro constant, and h is Planck constant.

The negative Gibbs energy indicates the spontaneity of the process, which rises with increasing temperature, increasing adsorption. The result indicates the stability of the reaction and (ΔG) decreases at higher temperatures. The analysis shows very low activation energy. The positive process enthalpy (3.43, 3.20, 3.47, and 3.27 kJ/mol) indicates that the reaction is endothermic. In entropy, this parameter value was negative, which suggests the randomness of adsorption [17].

The multitude of equations and descriptions as well as the ambiguity of the obtained results confirm the need for further research to understand the essence of the Fenton process conducted in a heterogeneous manner.

REFERENCES

1. Rahim Pouran, S., et al., Niobium substituted magnetite as a strong heterogeneous Fenton catalyst for wastewater treatment. *Applied Surface Science*, 2015. **351**: p. 175–187.
2. Zhou, H., et al., Removal of 2,4-dichlorophenol from contaminated soil by a heterogeneous ZVI/EDTA/Air Fenton-like system. *Separation and Purification Technology*, 2014. **132**: p. 346–353.

3. Wang, C., H. Liu, and Z. Sun, Heterogeneous photo-Fenton reaction catalyzed by nanosized iron oxides for water treatment. *International Journal of Photoenergy*, 2012. **2012**: p. 801694.
4. Ma, J., et al., Novel magnetic porous carbon spheres derived from chelating resin as a heterogeneous Fenton catalyst for the removal of methylene blue from aqueous solution. *Journal of Colloid and Interface Science*, 2015. **446**: p. 298–306.
5. Wu, Y., et al., Characteristics and mechanisms of kaolinite-supported zero-valent iron/H2O2 system for nitrobenzene degradation. *CLEAN – Soil, Air, Water*, 2017. **45**(3): p. 1600826.
6. Saratale, R.G., et al., Hydroxamic acid mediated heterogeneous Fenton-like catalysts for the efficient removal of Acid Red 88, textile wastewater and their phytotoxicity studies. *Ecotoxicol Environ Saf*, 2019. **167**: p. 385–395.
7. Kumar, V., et al., Degradation of mixed dye via heterogeneous Fenton process: Studies of calcination, toxicity evaluation, and kinetics. *Water Environ Res*, 2020. **92**(2): p. 211–221.
8. Shaban, Y.A., et al., Photocatalytic degradation of phenol in natural seawater using visible light active carbon modified (CM)-n-TiO2 nanoparticles under UV light and natural sunlight illuminations. *Chemosphere*, 2013. **91**(3): p. 307–313.
9. Kartal, O., M. Erol, and H. Oguz, Photocatalytic destruction of phenol by TiO2 powders. *Chemical Engineering & Technology - CHEM ENG TECHNOL*, 2001. **24**: p. 645–649.
10. Wang, L., et al., Removal of chlorpheniramine in a nanoscale zero-valent iron induced heterogeneous Fenton system: Influencing factors and degradation intermediates. *Chemical Engineering Journal*, 2016. **284**: p. 1058–1067.
11. Nakagawa, H., S. Takagi, and J. Maekawa, Fered-Fenton process for the degradation of 1,4-dioxane with an activated carbon electrode: A kinetic model including active radicals. *Chemical Engineering Journal*, 2016. **296**: p. 398–405.
12. Bogacki, J., et al., Magnetite, hematite and zero-valent iron as co-catalysts in advanced oxidation processes application for cosmetic wastewater treatment. *Catalysts*, 2021. **11**(1): p. 9.
13. Marcinowski, P., et al., Magnetite and hematite in advanced oxidation processes application for cosmetic wastewater treatment. *Processes*, 2020. **8**(11): p. 1343.
14. Hashemian, S. and M. Mirshamsi, Kinetic and thermodynamic of adsorption of 2-picoline by sawdust from aqueous solution. *Journal of Industrial and Engineering Chemistry*, 2012. **18**(6): p. 2010–2015.
15. Chen, J. and L. Zhu, Heterogeneous UV-Fenton catalytic degradation of dyestuff in water with hydroxyl-Fe pillared bentonite. *Catalysis Today*, 2007. **126**(3): p. 463–470.
16. Lin, S.-S. and M.D. Gurol, Catalytic decomposition of hydrogen peroxide on iron oxide: kinetics, mechanism, and implications. *Environmental Science & Technology*, 1998. **32**(10): p. 1417–1423.
17. Hussain, S., et al., Enhanced ibuprofen removal by heterogeneous-Fenton process over Cu/ZrO2 and Fe/ZrO2 catalysts. *Journal of Environmental Chemical Engineering*, 2020. **8**(1): p. 103586.

9 Toxicity of the Catalyst and Products Formed in the Process

During the heterogeneous Fenton process, a great amount of compounds is generated, which can be toxic to the environment. There is a possibility that after the treatment process, by-products with more toxic properties than parent compounds will appear [1]. Long-term use of iron oxides is friendly for the environment due to saving the natural resources in deposits. However, the lack of stability of the compound can result in its release into the environment. Katsnelson *et al.* [2] confirmed that the particle size may also affect the toxicity of the catalyst. Based on an equivalent mass dose, nanoparticles have more significant toxicity than microparticles due to the penetration of the particles into the cell and disruption of them. Then, nanoparticles cause mechanical damage and cause local inflammation. Inside the cells, nanoparticles agglomerate and disrupt the functions of organelles and membrane transport [3]. Moreover, nano-Fe_2O_3 can also be generated due to secondary formation [2]. Some nanoparticles are characterized by biocidal, biostatic, and even self-cleaning properties, which encourages researchers to use them in wastewater treatment despite the potential hazard [4].

Therefore, it is essential to constantly work on the stability of the catalyst used in the process and check if it poses any threat to living organisms. Studies are carried out on various types of organisms, such as bacteria, plants, and even higher animals like mammals. In the research performed on *Xanthobacter flavus* FB71, Wang *et al.* [5] noticed the inherence of toxic compounds, which can adversely affect the development after the process [5]. Büyüksönmez *et al.* [6] associated harmful effects with the production of hydroxyl and other oxygen-based radicals. The cells launched a protective mechanism against oxygen radicals, which increased catalase activity [6]. Iron oxides, commonly used as catalysts in the Fenton process, are considered to be nontoxic. Yet, it has been reported that Ceriodaphnia *dubia* accumulates nano-Fe_2O_3 as a result of ingestion. Additionally, the compound also serves as a carrier of As^{5+} and therefore increases toxicity [7].

The toxicity of the compounds against *Escherichia coli* and *Saccharomyces cerevisiae* was observed in the research performed by Schwegmann *et al.* [8]. They found that in the highest concentration of FeO_x, the survival rate was reduced to 1 and 25%, respectively. It also needs to be noted that the decrease in viability was dose-dependent, and the antibacterial effect was strictly related to the nanosize of particles [8].

Brunner *et al.* [9], demonstrated the cytotoxicity of the Fe_2O_3 to the rodent 3T3 fibroblast cells and human mesothelioma cell line (MSTO). They observed a cell-type specific response, as slower-growing 3T3 cells were only slightly affected by

DOI: 10.1201/9781003364085-9

the addition of up to 30 ppm iron oxide. However, 3.75 ppm of iron oxide reduced the cell parameters MTT-conversion and DNA content of faster proliferating MSTO cells. In addition, Pisanic *et al.* [10] noticed that the cytotoxicity of the same compound to the cell line derived from a pheochromocytoma of the rat adrenal medulla (PC12 cells). The magnetite and the other iron oxides used in nanoscale can oxidize, reducing pollutants' toxicity to mammalian cells. This research correlates the low neurotoxicity of cells with low redox activity [11].

In the heterogeneous Fenton process, contaminants are decomposed due to rapid oxidation in the presence of a catalyst [12]. In such a reaction, pollutants are transformed into simpler compounds that are usually less toxic [12,13]. However, as a consequence of incomplete mineralization also hazardous, potentially toxic compounds may occur in the effluent [12,14]. Therefore, great efforts and concerns need to be taken, as the toxicity of by-products also must be checked.

Cleveland *et al.* [15] evaluated the toxicity of effluent of degradation of bisphenol A (BPA) by carbon nanotube-supported Fe_3O_4 used in heterogeneous Fenton process against *E. coli*. In addition to the nearly complete removal of BPA and significant removal of COD, they observed almost no inhibition of *E. coli* in the presence of final reaction effluents [15].

GilPavas and Correa-Sánchez [16] performed acute toxicity tests with *Artemia salina* on textile wastewater treated with a heterogeneous electro-Fenton process assisted by scrap zero-valent iron. They noted that untreated wastewater caused a 100% mortality, which also remained in the samples treated for 30 and 60 min. Because biodegradability had increased significantly, they connected the high toxicity of effluent with the formation of harmful compounds. It is worth noting that after combination with activated carbon treatment, the toxicity was reduced to 10 and 20%, respectively [16]. In another approach, GilPavas and co-workers [17] once again compared the toxicity of textile wastewater against *A. salina*. This time they utilized scrap zero-valent iron as the replacement for dissolved iron in the Fenton process. Compared with samples of untreated wastewater, they observed a 67% decrease in mortality of *A. salina*. They also compared obtained results with the mortality of *A. salina* in a pure grow medium. This ensured that the transformation of the chemical composition of treated wastewater corresponds to a decrease in toxicity [17]. Lan *et al.* [18] carried out heterogeneous photodegradation of pentachlorophenol (PCP) with goethite, hematite, and oxalate under UVA illumination. To evaluate the toxicity, they utilized a Microtox analyzer, a standardized method of measuring acute toxicity against marine bacterium *Vibrio fischeri*. In terms of light loss, which is the basis of this approach, they confirmed the detoxification of PCP. For the samples (1.2 mM) after 1 hour of treatment, they observed a decrease in the ratios of light loss from 83 and 84% to 54 and 46% for goethite and hematite, respectively. Additionally, they checked qualitatively for intermediates from PCP degradation. Interestingly, more active substances could be generated in the case of hematite than goethite. This confirms that by-products were responsible for the slightly greater toxicity of effluent obtained with hematite as the catalyst [18].

A high increase in toxicity due to the degradation of tetracycline in an ultrasound-assisted heterogeneous Fenton process with magnetite as a catalyst was reported by Hou *et al.* [19]. In a 48-hour immobilization assay with *Daphnia magna*, they found a 60% mortality rate caused by a raw tetracycline solution. After 60 minutes of treatment, 100% of *D. magna* was immobilized. Observed toxicity was connected with by-products formed in tetracycline degradation, identified by liquid chromatography (LC) tandem mass spectrometry (MS). Interestingly, mentioned by-products exhibited high acute toxicity only at high concentrations, while no decayed toxicity was observed at low ones. Consequently, after 180 min of treatment, the mortality rate was only 55%, and therefore prolonging the time of reaction reduced the acute toxicity of *D. magna* [19].

The larger organisms against which scientists checked process safety are plants such as *Phaseolus mungo* and the mung bean. Saratale *et al.* [1] checked the phytotoxicity of dye effluents and products of the process ad used for it the *P. mungo*. The inhibition of seed and length germination and length of the plum roots were not noticed in the products of the process. A significant deterioration of germination capacity and reduction in the length of the rootlets in the tests carried out on the sludge, and the dye was however observed [1]. Kumar *et al.* [20] performed a toxicity test, in which they noticed a change in the germination efficiency of mung bean (*Virginia radiata*) and its root length. After seven days of the test, they noticed germination of seed growth of 0.2 cm for untreated wastewater. However, in the case of treated wastewater, the seed root has grown to 28 cm, which proves a decrease in toxicity due to the Fenton process [20].

Gonzalez-Gil *et al.* [21] performed phytotoxicity tests such as acute ecotoxicity. In terms of analysis, they performed mitochondrial activity studies, chronic biological tests, and quantification of chlorophyll on *Polystichum setiferum* spores. Ecotoxicological tests noticed the high efficiency of the process by reduction of the acute toxicity of river water samples. However, chronic toxicity, which was checked in the post-treatment samples, occurred. Other unanalyzed toxic substances, probably of inorganic nature and potentially harmful to plant development, were also released [21].

Chatzimarkou and Stalikas [22] carried out the degradation of estriol with magnetite and graphene oxide composites in the heterogeneous Fenton process. Consequently, they evaluated the toxicity of products of degradation against zebrafish larvae. They did not notice a difference in the lethality of zebrafish larvae due to the Fenton treatment of estriol. Moreover, malformation of larvae also did not occur [22].

The largest organisms used in toxicity tests were mammals such as rats. The acute toxicity of the degradation intermediates was determined utilizing the TEST software. Before the process, the LD_{50} value of about 1,650 mg/kg was considered to be toxic. Guo *et al.* [23] noticed a 50% lethal dose of the rat (LD_{50}) and developmental toxicity. After the treatment process, intermediates LD_{50} were described as very toxic and toxic. The LC_{50} was significantly lower for the intermediates than in the starting dye and therefore was described as a non-toxic development factor [23].

The critical element of every process is intermediates and by-product generation, which can be hazardous to the environment. Intermediates formed as by-products of the process can sometimes be toxic, as shown in Table 9.1. The intensity of the

TABLE 9.1
Selected Intermediates of Treatment during the Heterogenous Fenton Process

Processes	Compounds	Intermediates of the Purification Process	Reference
Combination of ultrasonic and Fenton processes in the presence of magnetite nanostructures prepared by high-energy planetary ball mill	Acid Blue 185	2-methylpropanoic acid, Decanoic acid, Benzenebutanoic acid, tert-butyldimethyl ester, Phthalazine, 1-phenyl, 1,2,3,4-Tetrahydroisoquinoline, 6,7,8-trimethoxy-1,2-dimethyl, 1,2-Benzenedicarboxylic acid, bis(2-methylpropyl) ester, 3-ethyl-5-(2-ethylbutyl)- Octadecane	[24]
Heterogeneous oxidation by the Fenton method in an aqueous solution studied on nanometric magnetite (Fe_3O_4) as a catalyst provided by a high-energy planet ball milling process	Ciprofloxacin	Acetamide, 4-(2-ethoxy-1-hydroxyallyl)phenol, 1-(tert-butyl)-3-methylbenzene, N-(4-(3-hydroxypyrrolidin-1-yl)but-2-yn-1-yl)-Nmethylacetamide, 1,2,3,4,5-pentamethylbenzen, -(2,6-dihydroxyphenyl)ethenone, 1,2,3,4,5,6-hexamethylbenzene, 5,8-dimethyl-1,2,3,4-tetrahydroacridin-9-amine,7-isopropyl-1,4-dimethylazulen, 3-diisopropylnaphthalen	[25]
Plant fluidized-bed heterogeneous Fenton process	Azo dye	Nitrobenzene, Phenol, Quinone, Propanedioic acid, 1,2-Benzenedicarboxylic Acid, propanamide, Pentanoic acid	[26]
Heterogeneous Fenton system with a nanoscale zero-valent iron	Chlorpheniramine	4-Hydroxy-4-methyl-2-pentanone, 2-Methylaminopyridine, 4-Chlorophenol, 2-Propionylpyridine, 2-Acetylpyridine, 4-Chlorophenyl-2-pyridyl ketone	[30]
Heterogeneous electro-Fenton process using Cu-doped Fe/Fe_2O_3	Tetracycline	$C_{21}H_{18}O_8N_2Me$, $C_{21}H_{21}O_6N_2$	[31]
Process catalyzed by magnetic porous carbon spheres derived from the chelating resin	Methylene blue	mono-chlorophenol, catechol, 2-chlorohydroquinone, iminodiacetic acid, oxalic acid, acetic acid, and formic acid	[32]
Heterogeneous Fenton with Fe-TiO_2 Visible	Atrazine	2-chloro-4-ethylamino-6-propanolamine-1,3,5-triazine; 2-Chloro-4-acetamido-6-isopropylamino-1,3,5-triazine; 2-hydroxy-4,6-dinitro-1,3,5-triazine; 2-chloro-4-amino-6-isopropylamino-1,3,5-triazine; 2-chloro-4-amino-6-nitro-1,3,5-triazine; 2-chloro-4-ethylamino-6-amino1,3,5-triazine; 2-hydroxy-4-amino-6-isopropylamino-1,3,5,	[33]

		-triazine; 2-chloro-4,6-dihydroxy-1,3,5-triazine; 2-chloro-4,6-diamino-1,3,5,-triazine; Cyanuric acid; Ammelide; 2-hydroxy-4,6-diamino-1,3,5-triazine	
Degradation of diazinon from aqueous solutions by electro-Fenton process	Diazinon	O,O-diethyl O-(2-isopropyl-6-methylpyrimidin-4-yl) phosphorothioate; O,O-diethyl O-(2-hydroxy-6-methylpyrimidin-4-yl) phosphorothioate; diethyl (2-isopropyl-6-methylpyrimidin-4-yl) phosphate; 2-isopropyl-6-methylpyrimidin-4-ol; O,O-diethyl O-hydrogen phosphorothioate; diethyl hydrogen phosphate; 2-isopropyl-4-methylpyrimidine; 4-methylpyrimidin-2-ol; 4-methylpyrimidine; (Z)-3-imino-N-methylprop-1-en-1-amine; ethane-1,2-diamine; ethyl methyl hydrogen phosphate; methyl dihydrogen phosphate; phosphoric acid; O-ethyl O-methyl O-hydrogen phosphorothioate; O-ethyl O,O-dihydrogen phosphorothioate; phosphorothioic O,O,O-acid	[34]
Pulsed discharge plasma with heterogeneous Fenton process using Fe_3O_4 nanocomposites	Chloramphenicol	(4-Nitrophenyl)methanediol, hydroxy(4-hydroxyphenyl) oxoammonium, 4-hydroxybenzoic acid,	[23]
Wastewater treatment with heterogeneous Fenton-type catalysts based on porous materials	EDTA	IMDA, glycine, ED3A, U-EDDA, S-EDDA, EDMA, EDA, glyoxalic acid	[35]
Heterogeneous Fenton process mediated by SiO_2-coated nano zero-valent iron	p-ASA organic arsenic contaminants	Analine, phenol, p-Benzoquinone, p-Aminophenol, p-Hydroquinone, Fumaric acid, Maleic acid, trans,trans-2,4-Hexadienedioic acid	[36]

process and the cost of technology selection and treatment depend on detected intermediate compounds. The type of intermediates depends on the kind of removed pollutant. Therefore, it is essential to identify and recognize by-products while determining the toxicity of the process. The most common methods of estimating the intermediates in the heterogeneous Fenton process utilize sophisticated techniques such as gas chromatography coupled with mass spectrometry (GC-MS) or high-performance liquid chromatography with a mass spectrometer (HPLC-MS). Selected results are shown in Table 1.1. Acisil *et al.* [24], noticed seven intermediates after the process, while Hassani *et al.* [25] identified ten compounds such as acetamide, 4-(2-ethoxy-1-hydroxyallyl), phenol, 1-(tert-butyl)-3-methylbenzene, and 1,2,3,4,5-pentamethylbenzene. Aghdasinia *et al.* [26] identified seven by-products, especially nitrobenzene, phenol, quinone, propanedioic acid, 1,2-benzenedicarboxylic acid, propenamide, and pentatonic, after azo dye degradation. Guo *et al.* [23] noticed the generation of the intermediates mentioned above by the reduction of the aromatic protein compounds and soluble microbial. They identified it with two methods: catalytic plasma and LC-MS. Firstly, the phenyl nitrile is reduced due to the weaker dissociation energy. In the next stage of the process, the amine group oxidation, hydroxyl addition, and carboxyl are substituted [27]. Also, it breaks down bonds in compounds such as CeN and CeC. The aromas broke the benzene ring and formation of organic (formic, acetic, and oxalic) and inorganic acids [28]. During the process, organic acids are oxidized. The final products of the process are neutral, harmless, natural compounds such as water and carbon dioxide. Hydrogen peroxide can react with aromatic organic by breaking the bond and adding or removing hydrogen to the original structure of the compound [29].

REFERENCES

1. Saratale, R.G., et al., Hydroxamic acid mediated heterogeneous Fenton-like catalysts for the efficient removal of Acid Red 88, textile wastewater and their phytotoxicity studies. *Ecotoxicol Environ Saf*, 2019. **167**: p. 385–395.
2. Katsnelson, B.A., et al., Subchronic systemic toxicity and bioaccumulation of Fe3O4 nano- and microparticles following repeated intraperitoneal administration to rats. *Int J Toxicol*, 2011. **30**(1): p. 59–68.
3. Karwowska, E., Antibacterial potential of nanocomposite-based materials - A short review. *Nanotechnology Reviews*, 2016. **6**.
4. Jakubczak, M., et al., Filtration materials modified with 2D nanocomposites—a new perspective for point-of-use water treatment. *Materials*, 2021. **14**(1): p. 182.
5. Wang, C., H. Liu, and Z. Sun, Heterogeneous photo-Fenton reaction catalyzed by nanosized iron oxides for water treatment. *International Journal of Photoenergy*, 2012. **2012**: p. 801694.
6. Buyuksonmez, F., et al., Toxic effects of modified fenton reactions on xanthobacter flavus FB71. *Appl Environ Microbiol*, 1998. **64**(10): p. 3759–3764.
7. Hu, J., et al., Bioaccumulation of Fe2O3(magnetic) nanoparticles in Ceriodaphnia dubia. *Environ Pollut*, 2012. **162**: p. 216–222.
8. Schwegmann, H., A.J. Feitz, and F.H. Frimmel, Influence of the zeta potential on the sorption and toxicity of iron oxide nanoparticles on S. cerevisiae and E. coli. *Journal of Colloid and Interface Science*, 2010. **347**(1): p. 43–48.

9. Brunner, T.J., et al., In vitro cytotoxicity of oxide nanoparticles: comparison to asbestos, silica, and the effect of particle solubility. *Environmental Science & Technology*, 2006. **40**(14): p. 4374–4381.
10. Pisanic, T.R., et al., Nanotoxicity of iron oxide nanoparticle internalization in growing neurons. *Biomaterials*, 2007. **28**(16): p. 2572–2581.
11. Phenrat, T., et al., Partial oxidation ("aging") and surface modification decrease the toxicity of nanosized zerovalent iron. *Environmental Science & Technology*, 2009. **43**(1): p. 195–200.
12. Jain, B., et al., Treatment of organic pollutants by homogeneous and heterogeneous Fenton reaction processes. *Environmental Chemistry Letters*, 2018. **16**(3): p. 947–967.
13. Baselt, R., Encyclopedia of toxicology. *Journal of Analytical Toxicology*, 2014. **38**(7): p. 464- 464.
14. Shen, J., T. Ding, and M. Zhang, *10 - Analytical techniques and challenges for removal of pharmaceuticals and personal care products in water*, in Pharmaceuticals and Personal Care Products: Waste Management and Treatment Technology, M.N.V. Prasad, M. Vithanage, and A. Kapley, Editors. 2019, Butterworth-Heinemann. p. 239–257.
15. Cleveland, V., J.-P. Bingham, and E. Kan, Heterogeneous Fenton degradation of bisphenol A by carbon nanotube-supported Fe3O4. *Separation and Purification Technology*, 2014. **133**: p. 388–395.
16. GilPavas, E. and S. Correa-Sánchez, Optimization of the heterogeneous electro-Fenton process assisted by scrap zero-valent iron for treating textile wastewater: Assessment of toxicity and biodegradability. *Journal of Water Process Engineering*, 2019. **32**: p. 100924.
17. GilPavas, E., S. Correa-Sánchez, and D.A. Acosta, Using scrap zero valent iron to replace dissolved iron in the Fenton process for textile wastewater treatment: Optimization and assessment of toxicity and biodegradability. *Environmental Pollution*, 2019. **252**: p. 1709–1718.
18. Lan, Q., et al., Heterogeneous photodegradation of pentachlorophenol and iron cycling with goethite, hematite and oxalate under UVA illumination. *J Hazard Mater*, 2010. **174**(1–3): p. 64–70.
19. Hou, L., et al., Ultrasound-assisted heterogeneous Fenton-like degradation of tetracycline over a magnetite catalyst. *Journal of Hazardous Materials*, 2016. **302**: p. 458–467.
20. Kumar, V., et al., Degradation of mixed dye via heterogeneous Fenton process: Studies of calcination, toxicity evaluation, and kinetics. *Water Environ Res*, 2020. **92**(2): p. 211–221.
21. Rodríguez-Gil, J.L., et al., Heterogeneous photo-Fenton treatment for the reduction of pharmaceutical contamination in Madrid rivers and ecotoxicological evaluation by a miniaturized fern spores bioassay. *Chemosphere*, 2010. **80**(4): p. 381–388.
22. Chatzimarkou, A. and C. Stalikas, Adsorptive removal of estriol from water using graphene-based materials and their magnetite composites: heterogeneous Fenton-like non-toxic degradation on magnetite/graphene oxide. *International Journal of Environmental Research*, 2020. **14**(3): p. 269–287.
23. Guo, H., et al., Degradation of chloramphenicol by pulsed discharge plasma with heterogeneous Fenton process using Fe3O4 nanocomposites. *Separation and Purification Technology*, 2020. **253**: p. 117540.
24. Acisli, O., et al., Combination of ultrasonic and Fenton processes in the presence of magnetite nanostructures prepared by high energy planetary ball mill. *Ultrasonics Sonochemistry*, 2017. **34**: p. 754–762.
25. Hassani, A., et al., Preparation of magnetite nanoparticles by high-energy planetary ball mill and its application for ciprofloxacin degradation through heterogeneous Fenton process. *J Environ Manage*, 2018. **211**: p. 53–62.

26. Aghdasinia, H., et al., Central composite design optimization of pilot plant fluidized-bed heterogeneous Fenton process for degradation of an azo dye. *Environmental Technology*, 2016. **37**(21): p. 2703–2712.
27. Chen, J., Y. Xia, and Q. Dai, Electrochemical degradation of chloramphenicol with a novel Al doped PbO2 electrode: Performance, kinetics and degradation mechanism. *Electrochimica Acta*, 2015. **165**: p. 277–287.
28. Garcia-Segura, S., E.B. Cavalcanti, and E. Brillas, Mineralization of the antibiotic chloramphenicol by solar photoelectro-Fenton: From stirred tank reactor to solar pre-pilot plant. *Applied Catalysis B: Environmental*, 2014. **144**: p. 588–598.
29. He, H. and Z. Zhou, Electro-Fenton process for water and wastewater treatment. *Critical Reviews in Environmental Science and Technology*, 2017. **47**: p. 1–32.
30. Wang, L., et al., Removal of chlorpheniramine in a nanoscale zero-valent iron induced heterogeneous Fenton system: Influencing factors and degradation intermediates. *Chemical Engineering Journal*, 2016. **284**: p. 1058–1067.
31. Luo, T., et al., Efficient degradation of tetracycline by heterogeneous electro-Fenton process using Cu-doped Fe@Fe2O3: Mechanism and degradation pathway. *Chemical Engineering Journal*, 2020. **382**: p. 122970.
32. Zhou, H., et al., Removal of 2,4-dichlorophenol from contaminated soil by a heterogeneous ZVI/EDTA/Air Fenton-like system. *Separation and Purification Technology*, 2014. **132**: p. 346–353.
33. Yang, N., et al., Study on the efficacy and mechanism of Fe-TiO(2) visible heterogeneous Fenton catalytic degradation of atrazine. *Chemosphere*, 2020. **252**: p. 126333.
34. Heidari, M., et al., Degradation of diazinon from aqueous solutions by electro-Fenton process: effect of operating parameters, intermediate identification, degradation pathway, and optimization using response surface methodology (RSM). *Separation Science and Technology*, 2020: p. 1–13.
35. Hartmann, M., S. Kullmann, and H. Keller, Wastewater treatment with heterogeneous Fenton-type catalysts based on porous materials. *Journal of Materials Chemistry*, 2010. **20**(41): p. 9002–9017.
36. Heidari, Z., et al., Application of mineral iron-based natural catalysts in electro-Fenton process: a comparative study. *Catalysts*, 2021. **11**(1): p. 57.

10 Reusability of Catalyst

The ecological aspect is one of the essential elements in every process, as shown in Figure 10.1.

Scientists are determined to develop methods that will be more friendly to the environment and do not generate additional pollution. The heterogeneous Fenton process is rather safe for the natural ecosystems and surrounding organisms of the trophic chain, and therefore it stands out from other methods. One of the advantages of the process is the possibility of the utilization of magnetic iron oxides. Such catalysts can be easily recovered and reused. The magnetic iron minerals can be separated by applying a magnetic field in the reactor, then separated and then re-dispersed in the solution for reuse [1,2]. Different research groups readily investigate the possibility of recycling the selected materials, which can be reused in the process. Moreover, as shown in Table 10.1, recovered catalysts are still effective.

Zhou *et al.* [5] utilized magnetic carbon composite material with good stability, reusability, and low use cost for seven cleaning cycles [5]. Guo *et al.* [3] carried out research in which the catalyst modified with graphene preserved its reactivity in five purification cycles without significantly losing its properties and effectiveness [3]. Eshaq *et al.* [19] used porous $FeVO_4$ nanorods decorated on CeO_2 nanocubes as the catalyst in six cleaning cycles, while Pouran *et al.* [20] noticed only four working cycles of the applied catalyst. During the processes, they did not notice a significant decrease in catalytic activity due to the stability of the applied catalyst and the low

FIGURE 10.1 Ecological aspect of heterogeneous Fenton process (created using BioRender.com).

DOI: 10.1201/9781003364085-10

TABLE 10.1
Popular Recycled Material Used in the Heterogenous Fenton Process (Self-prepared)

Recycled Material	Cycles of Purification	Effective	Reference
Catalyst with the addition of graphene	Five cycles	75.8% degradation Rhodamine	[3]
Modified iron-carbon	Eight cycles	33.8% degradation TOC	[4]
Magnetic carbon material	Seven cycles	Good stability	[5]
Magnetic porous carbon spheres derived from the chelating resin	Six purification cycles	80% MB and 40% degradation TOC	[6]
Magnetic	Four cycles	CAP remained declined to only 3.8%	[7]
Fe-SBA-15 nanocomposites	Three cycles	80% degradtins ACE	[8]
Iron Molybdophospahate (FeMoPO)	Three cycles	Removal efficiency reduced from 84.9, 65, 60 to 58%	[9]
4-nitrophenol	Six cycles	The sonophotocatalytic performance slightly decreased	[10]
Cu-doped Fe/Fe_2O_3	Seven cycles	Six cycles 100%; after seven cycles a significant decrease in effectiveness	[11]
ZnO nanoparticles	Six cycles	After six cycles degradation of 95%	[12]
CFZ antibiotic	Fifth cycles	After the fifth cycle degradation of 95%	[13]
Fe-PILC	Three cycles	After three cycles degradation of 90%	[14]
Titanomagnetite	Fifth cycles	The contents of element C on titanomagnetite after the fifth run was 1.72%, respectively	[15]
Iron-loaded black soil	Three cycles	Degradation achieved was 74.02% for 3rd runs of catalyst	[16]
Sewage sludge and iron sludge (the sewage sludge derived activated carbon-supported iron oxide)	Eight cycles	TOC abatement efficiency still reached 58.1% after eight cycle's utilization	[17]
HpFe-GAC	Ten cycles	88.1% and 72.0% of dyes in the first and 10th runs	[18]

leaching of Fe and Nb ions [20]. Zhang *et al.* [4] did the eighth cycle of recycling material and noticed the TOC removal of 33.8%. Ma *et al.* [6] used catalysts in six treatment cycles and they detected that the catalyst discolored 80% of methylene blue. Moreover, they also obtained 40% TOC removal efficiency [6].

Bolobajev *et al.* [21] reused ferric sludge in Fenton-based wastewater treatment. After the basic process, they collected supernatant and used concentrated sludge as an iron source for further treatment of landfill leachate, effluent from a plywood manufacturing plant, and leachate collected from a semicoke. They found that COD and DOC removal in the four treatment cycles with reused sludge remained almost identical. Moreover, the number of phenols and lignin with tannins was reduced by 98% [21]. Cao *et al.* [22] also employed iron-containing sludge as the catalyst in the multicycle Fenton-like oxidative treatment. To recover the catalyst, they dewatered and dried the raw post-process sludge, which was then baked at 400°C for 20 min. In the six treatment cycles observed nearly the same COD and BOD_5/COD removal, which suggests that sludge from the Fenton-like reactions can be easily recovered and reused [22]. Shahrifun *et al.* [23] also reused iron sludge, which was generated in the solar Fenton oxidation. By recycling the iron sludge in five cycles, they could decompose effectively palm oil mill secondary effluent. However, they pointed out an important aspect: the alternating amount of ferrous iron and total iron that remained in the sludge in every cycle [23]. Azmi *et al.* [24] used a different approach, as they prepared clay decorated with iron particles. Such catalyst was evaluated for decolorization of Acid Green 25 in three runs. They observed not only the high effectiveness of decolorization in all runs but also low iron leaching from iron clay [24].

These results confirm that the heterogeneous Fenton process favors saving raw materials by reusing the catalyst with high treatment efficiency in more than one cycle. Another ecological aspect of the heterogenous Fenton process is the utilization of agricultural by-products and waste materials as catalysts. This is the most effective method to save and reuse raw materials. The process also uses natural sources to acquire catalysts such as carbon material prepared from agricultural nutshells and iron, $K_2S_2O_8$ with the addition of persulfate [5]. Ramirez *et al.* [25] used carbon-based catalysts produced from olive stone and carbon aerogel with iron oxides. They developed them by carbonization of organic resorcinol–formaldehyde polymer. Then, the carbon surface was decorated with iron particles [25]. Another advantage of the heterogenous Fenton process is the utilization of scrap iron or steel as a catalyst. This enables saving natural iron from the environment. In industry, steel, iron-rich dust, steel shot, and furnace slag, materials for possible catalytical applications, are produced in large quantities. In 2020, only in the United States, about 82 million tons of iron and steel scrap were recycled in different ways.

The effectiveness of this method was confirmed by Ali *et al.* [26], who noticed 98% removal of dye with steel industry waste applied as catalysts. In addition, the electromagnetic properties of steel make easy separation of the catalyst after the end of the treatment process. However, this catalyst also has disadvantages, such as the leaching of iron and the deactivation of the catalyst in subsequent working cycles. Moreover, the X-ray photoelectron spectroscopy (XPS) analysis proved high catalyst stability [26]. Luo *et al.* [11] used the catalyst for more than six cycles. Moreover, utilized Fe could leach out of probing of catalysts and Fe^0 in reactive sites. It is the effect of the exchange of Cu^0 and the continuous deposition of iron oxide and copper oxide. In addition, the increase in the efficiency of removing contaminants due to the moderate content of the Cu element causes a reduction in the efficiency of catalysis [11].

Saber *et al.* [27] applied scrap iron powder in the Fenton process to treat a biorefractory petroleum refinery effluent. Over 83% of the COD removal within 90 min confirmed the beneficial aspects of the approach for the pre-treatment petroleum wastewater. Contrary to iron salts, the authors indicate a low cost of using scrap iron and the possibility of recycling the catalyst due to coagulation [27]. Liang *et al.* [28] used zero-valent scrap iron in the Fenton process as a part of the developed three-step sludge treatment. With a scrap iron size of 0.1 cm used in the pre-ultrasound-thermal-acid-washed method, they reduced the value of total organic carbon (TOC) and dewater the secondary settler of domestic sludge. Interestingly, the effectiveness of such a method was preserved for up to 15 cycles and was higher than classical Fenton-based processes. They also noticed additional effects like a decrease in toxicity and cost-effectiveness among other dewatering treatments [28]. Unlike mentioned scientists, Khajouei *et al.* [29] utilized scrap iron plates as electrodes in the electro-Fenton process to neutralize composting leachate. Under optimum conditions, they obtained 63.4, 78.7, and 99.3% reductions in COD, BOD_5, and PO_4–P, respectively, and were much higher than the ones for the standard Fenton method [29]. Chakinala *et al.* [30] applied a modified advanced Fenton process for industrial wastewater contaminated with phenolics with scrap iron as a catalyst. Additionally, they optimized parameters of the process such as pressure, pH, dilution of wastewater with fresh one, and H_2O_2 concentration. Interestingly, higher initial pressures in the reactor provided quicker TOC mineralization and low dissolved iron content. They also recognized advantages such as the abundance of scrap iron metal acting as a cheap catalyst [30]. Lan and Wu [31] applied scrap iron in the Fenton process supported by flocculation. They utilized Fe^{2+} ions to process wastes against one another. What is more, Fe^{3+} ions were linked up with the flocculation. Therefore, they were able to improve the overall effectiveness of the process by cutting down the costs [31].

Morikawa [32] developed iron-based catalysts using tea dregs and coffee grounds as raw materials in green synthesis. Both catalysts revealed superior activity in the model methylene blue dye degradation, as after 20 min, complete decolorization occurred. They also showed extraordinary biocidal activity against *E. coli* bacteria [32]. Xu *et al.* [33] also used iron-rich natural material to fabricate Biochar-supported iron nanocomposite for application in the heterogeneous Fenton process. For this purpose, they collected paddy rice root from the suburban area of Hangzhou city, which was then washed, dried, cut into pieces, and carbonized. They confirmed the presence of iron/iron oxide by both SEM and TEM.

Later, they studied its effectiveness in the degradation of brilliant red X-3B dye. They observed an increase in decolorization rate with the decreasing pH. At pH 3.0–4.0, the decolorization of investigated dye was the quickest. They also compared obtained catalyst with iron micron particles in a few working cycles. Iron micron particles' effectiveness began to decrease after three uses. However, for obtained catalyst could be reused even seven times [33].

REFERENCES

1. Li, X.Z., et al., Photocatalytic oxidation using a new catalyst--TiO2 microsphere--for water and wastewater treatment. *Environ Sci Technol*, 2003. **37**(17): p. 3989–3994.

2. Lai, B.-H., C.-C. Yeh, and D.-H. Chen, Surface modification of iron oxide nanoparticles with polyarginine as a highly positively charged magnetic nano-adsorbent for fast and effective recovery of acid proteins. *Process Biochemistry*, 2012. **47**(5): p. 799–805.
3. Guo, S., et al., Graphene modified iron sludge derived from homogeneous Fenton process as an efficient heterogeneous Fenton catalyst for degradation of organic pollutants. *Microporous and Mesoporous Materials*, 2017. **238**: p. 62–68.
4. Zhang, C., et al., Heterogeneous electro-Fenton using modified iron–carbon as catalyst for 2,4-dichlorophenol degradation: Influence factors, mechanism and degradation pathway. *Water Research*, 2015. **70**: p. 414–424.
5. Zhou, L., et al., Fabrication of magnetic carbon composites from peanut shells and its application as a heterogeneous Fenton catalyst in removal of methylene blue. *Applied Surface Science*, 2015. **324**: p. 490–498.
6. Ma, J., et al., Novel magnetic porous carbon spheres derived from chelating resin as a heterogeneous Fenton catalyst for the removal of methylene blue from aqueous solution. *Journal of Colloid and Interface Science*, 2015. **446**: p. 298–306.
7. Guo, H., et al., Degradation of chloramphenicol by pulsed discharge plasma with heterogeneous Fenton process using Fe3O4 nanocomposites. *Separation and Purification Technology*, 2020. **253**: p. 117540.
8. Benzaquén, T.B., et al., Heterogeneous Fenton reaction for the treatment of ACE in residual waters of pharmacological origin using Fe-SBA-15 nanocomposites. *Molecular Catalysis*, 2020. **481**: p. 110239.
9. Niveditha, S. and R. Gandhimathi, Mineralization of stabilized landfill leachate by heterogeneous Fenton process with RSM optimization. *Separation Science and Technology*, 2020: p. 1–10.
10. Garrido-Ramírez, E.G., et al., Characterization of nanostructured allophane clays and their use as support of iron species in a heterogeneous electro-Fenton system. *Applied Clay Science*, 2013. **86**: p. 153–161.
11. Luo, T., et al., Efficient degradation of tetracycline by heterogeneous electro-Fenton process using Cu-doped Fe@Fe2O3: Mechanism and degradation pathway. *Chemical Engineering Journal*, 2020. **382**: p. 122970.
12. Jain, B., et al., Treatment of pharmaceutical wastewater by heterogeneous Fenton process: an innovative approach. *Nanotechnology for Environmental Engineering*, 2020. **5**(2): p. 13.
13. Heidari, Z., et al., Application of mineral iron-based natural catalysts in electro-Fenton process: a comparative study. *Catalysts*, 2021. **11**(1): p. 57.
14. Bel Hadjltaief, H., et al., Influence of operational parameters in the heterogeneous photo-Fenton discoloration of wastewaters in the presence of an iron-pillared Clay. *Industrial & Engineering Chemistry Research*, 2013. **52**(47): p. 16656–16665.
15. Yang, S., et al., Degradation of methylene blue by heterogeneous Fenton reaction using titanomagnetite at neutral pH values: process and affecting factors. *Industrial & Engineering Chemistry Research*, 2009. **48**(22): p. 9915–9921.
16. Kumar, V., et al., Degradation of mixed dye via heterogeneous Fenton process: Studies of calcination, toxicity evaluation, and kinetics. *Water Environ Res*, 2020. **92**(2): p. 211–221.
17. Munoz, M., et al., Preparation of magnetite-based catalysts and their application in heterogeneous Fenton oxidation – A review. *Applied Catalysis B: Environmental*, 2015. **176–177**: p. 249–265.
18. Saratale, R.G., et al., Hydroxamic acid mediated heterogeneous Fenton-like catalysts for the efficient removal of Acid Red 88, textile wastewater and their phytotoxicity studies. *Ecotoxicol Environ Saf*, 2019. **167**: p. 385–395.

19. Eshaq, G., et al., Superior performance of FeVO(4)@CeO(2) uniform core-shell nanostructures in heterogeneous Fenton-sonophotocatalytic degradation of 4-nitrophenol. *J Hazard Mater*, 2020. **382**: p. 121059.
20. Rahim Pouran, S., et al., Niobium substituted magnetite as a strong heterogeneous Fenton catalyst for wastewater treatment. *Applied Surface Science*, 2015. **351**: p. 175–187.
21. Bolobajev, J., et al., Reuse of ferric sludge as an iron source for the Fenton-based process in wastewater treatment. *Chemical Engineering Journal*, 2014. **255**: p. 8–13.
22. Cao, G.-m., et al., Regeneration and reuse of iron catalyst for Fenton-like reactions. *Journal of Hazardous Materials*, 2009. **172**(2): p. 1446–1449.
23. Ahmad Shahrifun, N.S., et al., Reusability of Fenton sludge to reduce cod and color on palm oil mill secondary effluent (POMSE). *Advanced Materials Research*, 2015. **1113**: p. 486–491.
24. Azmi, N.H.M., V. Vadivelu, and B. Hameed, Iron-clay as a reusable heterogeneous Fenton-like catalyst for decolorization of Acid Green 25. *Desalination and Water Treatment*, 2014. **52**.
25. Ramirez, J.H., et al., Azo-dye Orange II degradation by heterogeneous Fenton-like reaction using carbon-Fe catalysts. *Applied Catalysis B: Environmental*, 2007. **75**(3): p. 312–323.
26. Ali, M.E.M., T.A. Gad-Allah, and M.I. Badawy, Heterogeneous Fenton process using steel industry wastes for methyl orange degradation. *Applied Water Science*, 2013. **3**(1): p. 263–270.
27. Saber, A., et al., Optimization of Fenton-based treatment of petroleum refinery wastewater with scrap iron using response surface methodology. *Applied Water Science*, 2014. **4**(3): p. 283–290.
28. Liang, J., et al., High-level waste activated sludge dewaterability using Fenton-like process based on pretreated zero valent scrap iron as an in-situ cycle iron donator. *Journal of Hazardous Materials*, 2020. **391**: p. 122219.
29. Khajouei, G., et al., Treatment of composting leachate using electro-Fenton process with scrap iron plates as electrodes. *International Journal of Environmental Science and Technology*, 2018. **16**.
30. Chakinala, A.G., et al., A modified advanced Fenton process for industrial wastewater treatment. *Water Sci Technol*, 2007. **55**(12): p. 59–65.
31. Lan, S. and X. Wu, Influence of pH on treatment of medium pulping wastewater with micro electrolysis coupling with Fenton oxidation-flocculation technology. *Advanced Materials Research*, 2011. **233–235**: p. 1794–1798.
32. Morikawa, C.K., A new green approach to Fenton's chemistry using tea dregs and coffee grounds as raw material. *Green Processing and Synthesis*, 2014. **3**.
33. Xu, L., et al., Green and simple method for preparing iron oxide nanoparticles supported on mesoporous biochar as a Fenton catalyst. *Applied Organometallic Chemistry*, 2020. **34**(9): p. e5786.

11 Economical Aspects and Hydrogen Production

The expected price of the process is difficult to estimate due to the variable cost of applied water and electricity for various countries in the world. The nanomaterials are also difficult and costly to produce. For this reason, in the heterogeneous Fenton process in research, only small amounts of compounds are used. In addition, nanomaterials are good compounds to oxidize and passivate, due to their high surface area. Moreover, small doses of compounds, such as a few micrograms, can be an effective catalyst for the process. The costs of the iron powder are (average diameter 25 nm) $18/gram in the USA in 2020 [1,2]. However, costs increase when used in industry due to investments related to the construction of the installation, obtaining the relevant permits and technology installations. Barndok *et al.* [3] estimated the costs of the process based on operating parameters such as energy, chemicals, maintenance, analytical monitoring, and operating hours. They determined the energy consumption by considering the industrial price of electricity in Spain in 2017 (0.1 €/kWh) [3]. Mahamuni *et al.* [4] used the additional energy during pumping and cooling of the reagent. Also, it is necessary to use the equipment essential to power the propulsion [5].

To calculate the chemical costs industrialists used the annual cost indicators for chemical products [3]. Alibaba *et al.* [6] described the cost of reagents as 0.45 €/kg per H_2O_2 (50%), 0.3 €/kg for NaOH, 0.09 €/kg for Na_2SO_4 and 5 €/kg for Fe^0 powder. Munoz *et al.* [5] made the calculations by assuming a daily recovery of 90% of the catalyst, while the cost of storage for waste catalyst was 0.07 €/kg and the individual cost of using the process in industry is calculated with the price of construction of materials and equipment, as well as operating costs [5]. For example, in wastewater treatment the calculations are made for m^3 of treated wastewater [7]. Carrara *et al.* [8] assumed the value of 800 €/m^2 in their research. Alalm *et al.* [7] obtained an amortization cost of 1.52 €/m^3, while the cost of wastewater treatment using optimal conditions was 2.86 €/m^3.

According to research by Molinos-Senante *et al.* [9], the cost of maintenance was about 2% of the total amount spent on the investment. Alalm *et al.* [7], calculated costs of the reagents by the concentration of the product (kg/m^3) and the unit price €/kg). The prices were estimated according to the local rates. Also, the energy costs (EC, €/m^3) were calculated based on the power required to pump the wastewater (the unit price for energy was 0.12 €/kWh). Moreover, a slight reduction of hydrogen peroxide (from 45 to 30 mmol/L) low cost of process (2.54 €/m^3) and obtain 100% treatment efficiency [7].

Another method to decrease the prices is the optimization of the process parameters by applying the special statistical analysis. Statistical and mathematical methods

DOI: 10.1201/9781003364085-11

are used to optimize the process and reduce the costs of treatment. Aghdasinia *et al.* [10] used the central composite design (CCD) and the Design Expert, analyzing the research results. Niveditha *et al.* [11] used response surface methodology (RSM), based on Box Behnken Design (BBD) for analysis of variables parameters and generated a mathematical model equation [12]. Marcinowski *et al.* [13] used the analysis of variance (ANOVA) to optimize the process parameters.

Due to the use of statistical methods, in the future, scientists will be able to develop optimal doses of catalysts and compounds used in the process for the industry, as well as select appropriate process conditions. This method will result in increase in the efficiency of the process and a reduction of its costs, which will encourage the implementation of the technology in the industry.

What's more, the optimalization of the process is one the most important part of the method. To minimalize the cost of the process, scientists analyzed the various proportion of the Fenton process 'parameters to obtain the cheapest form of the method. Rodriguez *et al.* [14] decided to minimalize the cost of the process analyzed the effect of the light flux. They checked the efficiency of the process between 2.1 and 150 W, which is equivalent to 7 and 500 W/m^2 with using the amount of reagent allowed to efficiency decomposition pollutant. Interestingly, scientists observed the light flux could be less then 500 W/m^2, for H_2O_2 doses 6.5–20 g/L. However, to obtain effectiveness process, they did not have to use the amount of H_2O_2 in range 2.5–5.0 g/L. After the process this parameters, the analysis of the wastewater showed the COD of 250 mg O_2/L. Next, Rodriguez *et al.* checked the efficiency of the Fenton process for various light flux and 6.5, 10, and 20 g/L of H_2O_2. Scientists noticed excellent removal of color wastewater in all parameters in prepared research. To obtain the total decolourizations of the wastewater, they need only 1–5 min. Moreover, the efficiency did not depends of radiation flux. Interestingly, the independents of the radiation flux and amount of H_2O_2, the decomposition of DOC is noticed after first 15 min of the process. Next, the velocity of the decomposition of DOC decrease to 60 min and finally stop. Another situation was observed for the organic compounds. Scientists observed that the removal of organic compound is more efficiency with the higher radiation flux and amount of H_2O_2. Additionally, Rodriguez *et al.* observed that increasing the amount of oxidant case the lower light flux needed to efficiency decomposition wastewater. Additionally, Rodriguez *et al.* observed that increasing the amount of oxidant case the lower light flux needed to efficiency decomposition wastewater. What important, there is also a light flux, which is too small to decomposition wastewater. Thus, scientists should find a proportion between the amount of oxidant and light flux. It is important to optimalization process and thus, the cost of the method.

During the using the Fenton process as a method of decomposition industrial wastewater, the total cost of the process is sum of the capital, operating, and maintenance costs. Another important parameters which decide about the price of the system is the flow rate of the effluent, the nature of the wastewater, and the configuration of the reactors, among other issues. Rodriguez *et al.* analyzed the cost of the reagents and the cost of energy. The cost of the process was obtained by sum the price of hydrogen peroxide, ferrous sulphate, and energy consumption, as shown by Eq. (11.1). The additional costs were calculated using Eqs. (11.2–11.4).

$$operating\ \ cost = cost_{H_2O_2} + cost_{Fe^{2+}} + cost_{energy} \tag{11.1}$$

$$cost_{H_2O_2} = \frac{price\, H_2\left(\frac{€}{ton}\right)[H_2O_2] \times 10^{-3}\frac{ton}{m^3} \times \rho_{H_2O_2}\frac{kg}{L}}{\frac{\%\, H_{2O_2}}{100}\left(\frac{kg}{L}\right)} \tag{11.2}$$

$$cost_{Fe^{2+}} = \frac{\left[price\ \ FeSO_4{\cdot}7H_2O\left(\frac{€}{ton}\right)[Fe^{2+}] \times 10^{-6}\left(\frac{ton}{m^3}\right) \times \frac{MM_{FeSO_4} \cdot 7H_2O}{MMFe^{Fe2+}}\right]}{[\%\ \ of\ \ purity/100]} \tag{11.3}$$

$$cost_{energy} = \frac{\left[power\ \ of\ \ lamp \times 10^{-3}(kW) \times price\ \ energy\left(\frac{€}{kWh}\right) \times time_{reaction\,(h)}\right]}{[volume_{treated_{effluent}}\ (m^3)]} \tag{11.4}$$

Rodriguez *et al.* noticed the following cost of reagents and energy from Quimitecnica S.A.,: H_2O_2 (49.5% w/v, density at 25°C = 1.2 g/cm^3), 365 €/ton; $FeSO_4{\cdot}7H_2O$ (93% of purity), 233.7 €/ton, as well as 0.10 €/(kWh) (Tables 11.1 and 11.2).

TABLE 11.1
Vales of COD, BOD$_5$, and Visible Color (1:40 dilution) after 60 min of Oxidation for Different Doses of H_2O_2 and Light Flux Using the TQ150 Lamp (in Bold: Conditions That Allow Meeting the Discharge Limits at the Minimum Cost)

$[H_2O_2]$ (g/L)	Light Flux (W/m^2)	COD (mg O_2/L)	BOD$_5$ (mg O_2/L)	Visible Color after Dilution 1:40	Total Costs (€/m^3)
Acrylic					
2.5	500	342.3	25.3	not visible	21.0
5.0	500	271.4	27.2	not visible	23.3
6.5	253	311.1	25.8	not visible	15.5
6.5	500	227.2	33.3	not visible	24.7
10.0	7	300.7	24.1	not visible	9.4
10.0	107	278.7	39.7	not visible	13.2
10.0	220	246.9	42.6	not visible	17.4
10.0	253	237.2	44.0	not visible	18.7

(Continued)

TABLE 11.1 (Continued)
Vales of COD, BOD_5, and Visible Color (1:40 dilution) after 60 min of Oxidation for Different Doses of H_2O_2 and Light Flux Using the TQ150 Lamp (in Bold: Conditions That Allow Meeting the Discharge Limits at the Minimum Cost)

$[H_2O_2]$ (g/L)	Light Flux (W/m^2)	COD (mg O_2/L)	BOD_5 (mg O_2/L)	Visible Color after Dilution 1:40	Total Costs (€/m^3)
10.0	500	139.4	39.8	not visible	27.9
20.0	7	222.5	25.9	not visible	18.6
20.0	107	205.4	31.8	not visible	22.3
20.0	253	149.1	41.4	not visible	27.8
20.0	500	90.5	55.3	not visible	37.0
Cotton					
2.5	500	271.6	77.4	not visible	2.6
3.75	253	284.0	74.5	not visible	13.0
3.75	500	235.7	78.5	not visible	22.2
5.0	7	284.0	65.0	not visible	4.9
5.0	107	271.6	68.7	not visible	8.7
5.0	220	223.3	73.4	not visible	12.9
5.0	253	210.0	76.4	not visible	14.2
5.0	500	185.2	81.3	not visible	23.4
10.0	7	234.6	76.5	not visible	9.6
10.0	107	210.0	78.5	not visible	13.3
10.0	253	160.5	80.2	not visible	18.8
10.0	500	135.8	85.5	not visible	28.1
Polyester					
0.625	500	264.1	74.9	not visible	19.4
0.938	253	267.3	77.8	not visible	10.5
0.938	500	225.2	65.7	not visible	19.8
1.25	7	251.2	85.1	not visible	1.6
1.25	107	241.4	67.5	not visible	5.3
1.25	220	230.2	66.6	not visible	9.6
1.25	253	221.7	73.8	not visible	10.8
1.25	500	202.9	61.1	not visible	20.1
2.5	7	219.2	62.3	not visible	2.9
2.5	107	199.5	59.4	not visible	6.7
2.5	253	179.8	57.2	not visible	12.2
2.5	500	154.0	56.2	not visible	21.4

Source: https://pubs.acs.org/doi/full/10.1021/ie401301h?casa_token=xuKw0tvxYwAAAAAA%3Ak4uxnSTlQO_XG6fDTnxidr123cyEWOWFbZjaKMcqPgiY2Y-BNNbi3f_22xAfdzBXmYJwOS7Q5wOTbNs

TABLE 11.2
Vales of COD, BOD_5, and Visible Color (1:40 dilution) after 60 min of Oxidation for Different Doses of H_2O_2 and Light Flux Using Simulated Solar Radiation (in Bold: Conditions That Allow Meeting the Discharge Limits at the Minimum Cost)

$[H_2O_2]$ (g/L)	Light Flux (W/m^2)	COD (mg O_2/L)	BOD_5 (mg O_2/L)	Visible Color after Dilution 1:40	Total Costs (€/m^3)
Acrylic					
6.5	**500**	**232.1**	**32.4**	**not visible**	**5.8**
10.0	253	242.0	44.6	not visible	9.2
10.0	500	143.2	33.1	not visible	9.2
20.0	253	150.6	40.8	not visible	18.3
20.	500	101.2	55.8	not visible	
Cotton					
3.75	**500**	**244.5**	**76.3**	not visible	**3.5**
5.0	253	207.8	75.0	not visible	4.7
5.0	500	183.4	80.9	not visible	4.7
10.0	253	158.9	79.3	not visible	9.3
10.0	500	134.5	83.5	not visible	9.3
Polyester					
0.938	**500**	**231.0**	**66.4**	**not visible**	**1.0**
1.25	253	221.1	74.3	not visible	1.3
1.25	500	206.4	61.4	not visible	1.3
2.5	253	179.4	60.8	not visible	2.7
2.5	500	167.1	57.6	not visible	2.7

Source: https://pubs.acs.org/doi/full/10.1021/ie401301h?casa_token=xuKw0tvxYwAAAAAA%3Ak4uxnSTlQO_XG6fDTnxidr123cyEWOWFbZjaKMcqPgiY2Y-BNNbi3f_22xAfdzBXmYJwOS7Q5wOTbNs

HYDROGEN PRODUCTION

The Fenton-iron process is a widely used advanced oxidation process that involves the generation of highly reactive hydroxyl radicals to oxidize and degrade organic pollutants in wastewater. In this process, iron acts as a catalyst to generate hydroxyl radicals from hydrogen peroxide (H_2O_2). However, hydrogen (H_2) can also play a role in the Fenton-iron process. Hydrogen is a by-product of the Fenton-iron process, produced from the reduction of H_2O_2 by Fe^{2+}. The reduction reaction of H_2O_2 by Fe^{2+} generates hydroxyl radicals, which are highly reactive and can degrade organic pollutants in wastewater. However, if excess H_2O_2 is added to the reaction, it can react with Fe^{2+} to form Fe^{3+} and hydroxide (OH^-) ions. This reaction consumes H_2O_2 and reduces the generation of hydroxyl radicals. Therefore, the presence of excess H_2 can help to scavenge excess H_2O_2 in the Fenton-iron process, thus promoting the generation of hydroxyl radicals. The H_2 can also react with hydroxyl radicals to form water, which is a benign end product of the reaction. This can prevent the formation of other unwanted byproducts that may be harmful to the environment. Special reactor is not required to obtain hydrogen during the Fenton-iron process. However, the amount of hydrogen produced depends on the concentration of H_2O_2, the type of iron catalyst used, and the pH of the reaction mixture. In some cases, hydrogen gas may accumulate in the reaction vessel during the Fenton-iron process. This can lead to the formation of an explosive mixture if the concentration of hydrogen exceeds the lower explosive limit (LEL). Therefore, it is important to ensure adequate ventilation and safety measures when carrying out the Fenton-iron process. In summary, while a special reactor is not required to obtain hydrogen during the Fenton-iron process, it is important to consider the potential hazards associated with the accumulation of hydrogen gas in the reaction vessel. Proper safety measures and precautions should be taken to ensure a safe and effective process [15,16].

Hydrogen production during the Fenton process is beneficial for several reasons. First, the production of hydrogen gas (H_2) can act as a scavenger for $^{\bullet}OH$, which can limit the formation of undesirable by-products, such as chlorinated organic compounds, during the treatment of certain types of pollutants. This can enhance the efficiency and selectivity of the Fenton process. Second, the production of hydrogen gas can be harnessed for various energy applications. Hydrogen is a clean and renewable fuel that can be used for the production of electricity, heating, and transportation. The use of hydrogen as a fuel can help reduce greenhouse gas emissions and promote sustainable energy practices. Finally, the production of hydrogen gas during the Fenton process can be an indicator of the effectiveness of the process. The rate of hydrogen gas production can be used as a measure of the rate of reaction and can be used to optimize the conditions for maximum efficiency. However, it is important to note that the production of hydrogen gas during the Fenton process can also present some challenges. The accumulation of hydrogen gas can lead to the formation of gas pockets and inhibit the reaction, which can limit the efficiency of the process. In addition, the presence of other gases in the system, such as oxygen, can affect the rate and selectivity of hydrogen gas production. Overall, the production of hydrogen gas during the Fenton process can offer several benefits, including enhanced efficiency and selectivity, sustainable energy production, and a measure of the effectiveness of the process. However, the production of hydrogen gas also requires careful consideration of the environmental conditions and the presence of other gases in the system [17–20].

To conclude, hydrogen plays a significant role in the Fenton-iron process by scavenging excess H_2O_2 and promoting the generation of hydroxyl radicals. It can also help to prevent the formation of unwanted byproducts, making the Fenton-iron process an effective and environmentally friendly method for the treatment of wastewater.

REFERENCES

1. Store, N., *Nanomaterials* [cited 2021 22 June]; Available from: https://www.nanomaterialstore.com/index.php.

2. Sigma-Aldrich., *Monomers*. 2021 [cited 2021 22 June]; Available from: https://www.sigmaaldrich.com/materials-science/material-science-products.html?TablePage=119470654&fbclid=IwAR2wzMH64Q6_Xj5GFotofPZxIPwIrgACIdzCks_DpXERrmqQnp0pL86GeNQ
3. Barndok, H., *Advanced oxidation processes for the treatment of industrial wastewater containing 1,4-dioxane*, in *Facultad de Ciencias Químicas Departamento de Ingeniería Química*. 2017, Universidad Complutense de Madrid: Madrid.
4. Mahamuni, N.N. and Y.G. Adewuyi, Advanced oxidation processes (AOPs) involving ultrasound for waste water treatment: A review with emphasis on cost estimation. *Ultrasonics Sonochemistry*, 2010. **17**(6): p. 990–1003.
5. Muñoz, I., *Life Cycle Assessment as a Tool for Green Chemistry: Application to Different Advanced Oxidation Processes for Wastewater Treatment*. 2006, Autonomous University of Barcelona: Spain.
6. Alibaba. *Alibaba Chemicals Market*. 2021 [cited 2021 22 June]; Available from: www.alibaba.com.
7. Gar Alalm, M., A. Tawfik, and S. Ookawara, Investigation of optimum conditions and costs estimation for degradation of phenol by solar photo-Fenton process. *Applied Water Science*, 2017. **7**(1): p. 375–382.
8. Carra, I., et al., Cost analysis of different hydrogen peroxide supply strategies in the solar photo-Fenton process. *Chemical Engineering Journal*, 2013. **224**: p. 75–81.
9. Molinos-Senante, M., F. Hernández-Sancho, and R. Sala-Garrido, Economic feasibility study for wastewater treatment: a cost-benefit analysis. *Sci Total Environ*, 2010. **408**(20): p. 4396–4402.
10. Vilardi, G., et al., Heterogeneous nZVI-induced Fenton oxidation process to enhance biodegradability of excavation by-products. *Chemical Engineering Journal*, 2018. **335**: p. 309–320.
11. Niveditha, S. and R. Gandhimathi, Mineralization of stabilized landfill leachate by heterogeneous Fenton process with RSM optimization. *Separation Science and Technology*, 2020: p. 1–10.
12. Tripathi, P., V.C. Srivastava, and A. Kumar, Optimization of an azo dye batch adsorption parameters using Box–Behnken design. *Desalination*, 2009. **249**(3): p. 1273–1279.
13. Bogacki, J., et al., Magnetite, hematite and zero-valent iron as co-catalysts in advanced oxidation processes application for cosmetic wastewater treatment. *Catalysts*, 2021. **11**(1): p. 9.
14. Rodrigues, C.S.D., L.M. Madeira, and R.A.R. Boaventura, Optimization and economic analysis of textile wastewater treatment by photo-Fenton process under artificial and simulated solar radiation. *Industrial & Engineering Chemistry Research*, 2013. **52**(37): p. 13313–13324.
15. Nawaz, S., et al., Ultrasound-assisted hydrogen peroxide and iron sulfate mediated Fenton process as an efficient advanced oxidation process for the removal of Congo red dye. *Polish Journal of Environmental Studies*, 2022. **31**(3): p. 2749–2761.
16. Tokumura, M., R. Morito, and Y. Kawase, Photo-Fenton process for simultaneous colored wastewater treatment and electricity and hydrogen production. *Chemical Engineering Journal*, 2013. **221**: p. 81–89.
17. Yang, L., et al., Effective green electro-Fenton process induced by atomic hydrogen for rapid oxidation of organic pollutants over a highly active and reusable carbon based palladium nanocatalyst. *Applied Surface Science*, 2022. **602**: p. 154325.
18. Goi, A. and M. Trapido, Hydrogen peroxide photolysis, Fenton reagent and photo-Fenton for the degradation of nitrophenols: a comparative study. *Chemosphere*, 2002. **46**(6): p. 913–922.

19. Petrucci, E., A. Da Pozzo, and L. Di Palma, On the ability to electrogenerate hydrogen peroxide and to regenerate ferrous ions of three selected carbon-based cathodes for electro-Fenton processes. *Chemical Engineering Journal*, 2016. **283**: p. 750–758.
20. Iervolino, G., et al., Hydrogen production from glucose degradation in water and wastewater treated by Ru-LaFeO3/Fe2O3 magnetic particles photocatalysis and heterogeneous photo-Fenton. *International Journal of Hydrogen Energy*, 2018. **43**(4): p. 2184–2196.

12 Perspective and Summary

Although the Fenton process has been known for over 100 years, intensive research is still underway to fully understand its mechanism, the possibility of applying the process in practice and the possibility of process improvement. Thousands of research and theoretical publications have been written about the Fenton process and its various modifications. They have been also summarized in hundreds of review articles. The research is carried out both with the use of model pollutants as well as synthetic and real wastewater. The influence of reagent doses, process time, pH, temperature, matrix parameters, or type of catalyst on the effectiveness of the process is investigated. In many cases, it is possible to eliminate model pollutants or achieve a degree of their conversion to carbon dioxide and water close to 100%. However, in many situations, especially in the case of heavily polluted wastewater, only partial conversion is achieved, and a certain number of pollutants remains in the treated wastewater. In the previous chapters, the chemistry of the Fenton process was analyzed, the catalysts that could be used were determined, as well as the optimal conditions for the process, and the reactors potentially used were described. The economic conditions of the applicability of the process were also considered. This allows us to determine the degree of complexity of the process, the need for its deeper study, and the huge application potential. The Fenton process itself, although from a technological point of view is simply feasible because the reagents, divalent iron ions, and hydrogen peroxide, are cheap and easily available, the process takes place under normal conditions, does not require a significant amount of energy, etc., has numerous disadvantages, including the required very acidic pH, significant salinity of wastewater and formation of post-process sludge. One of the solutions to these problems is the implementation of the process using heterogeneous catalysts. This solves the problem of post-process sludge, due to the relative ease of separating the catalyst, and at the same time allows the process to be carried out in conditions closer to neutral.

HFP is an AOP with great potential for industrial applications. Difficult-to-remove compounds are found in many industries and their scope is constantly expanding. The method is already very popular in industries such as textiles, pharmaceuticals, chemicals, cosmetics, and during projectile dewatering. However, as technology advances and the process is adapted for use in factories, this list may grow longer. Moreover, the method shows high efficiency in removing micropollutants resistant to biodegradation. Few methods offer such high efficiency in removing difficult-to-remove compounds. Therefore, this method should be popularized and developed. HFP can be combined with other processes to increase treatment efficiency.

DOI: 10.1201/9781003364085-12

This is one of the directions of the method development worth testing in factories and wastewater treatment plants.

- The implementation of technology requires the preparation of appropriate procedures and the transfer of the laboratory scale to an industrial scale. However, due to its numerous advantages, the process can be readily used, even by small points producing hardly removable pollutants, dangerous for the environment. Works related to the development of the technology used on a commercial scale are still in progress. However, it is currently difficult to determine the exact parameters of the process due to the wide variety of process modifications and the adaptation of parameters to specific conditions. The advantage of the method is low process costs and a high degree of treatment, but it would require research and design on a technological scale, which is not popular yet.
- New HFP catalysts should be developed. Crucial parameters that had to be studied and determined for new catalysts are among others leaching rate, stability, catalytic activity, or preparation costs. New catalysts include typically used iron-based ones as well as other metals. Research directions include the use of carriers on which monometallic, and bimetallic catalysts in the form of alloys, sinters, metal-organic frameworks, and others. During the performance of a heterogeneous process crucial aspect is the type of catalyst used, as well as increasing its efficiency. For the use of the catalyst in industry, the price of the compound should be reduced and its availability on the market increased. In addition, it is worth using catalyst forms that are convenient to use for various industries. It is necessary to adjust the appropriate dose, form, and dosing method of the compound. In this case, it is worth using waste iron as a catalyst, which can be cheap and effective. Additionally, it is one of the methods of metal waste management. An interesting solution is also the modification of the catalyst with various compounds that increase the efficiency of the process and the use of nanostructures.
- It is also important to constantly research the safety of the method for aquatic organisms and the entire environment. Before being implemented in the industry, the method must be tested well tested, so it is worth continuing research towards comprehensive method recognition. The process can be modified in many ways and improved, so it is essential to constantly increase its efficiency. During the preparation of the initial implementation project process for industrial plants, it is necessary to carry out wastewater from industrial plants.
- It is extremely important to describe precisely the mechanism of toxic organic pollutants by Fenton processes. As much existing research lacks kinetics determination, more detailed studies should be focused on the organic pollutants' degradation kinetics, the development of models simulating degradation mechanisms, and the identification of process intermediates.
- Economic analysis is necessary to move from the scale of beaker or vial laboratory tests to the semi-technical and industrial scale. Economic

aspects are crucial factors for the industrial scale development of Fenton processes. The detailed consideration of factors such as chemicals costs, reactor investment costs, and energy consumption costs should be accurately taken into account.
- The Fenton process is hardly unlikely to be the only treatment process. Due to the stringent legal requirements that must be met before the discharge of the treated wastewater to the receiver, a treatment technology consisting of at least several unit processes should be expected. The coupling of the Fenton process with other treatment processes, including biological treatment should be studied. Therefore, the purpose of using the Fenton process should be precisely defined; it is not the maximum reduction of the content of organic compounds after the process, but the maximum reduction of the toxicity of the compounds remaining after the process and the maximum increase in the susceptibility to biological treatment of pretreated wastewater.

A SHORT HISTORY OF THE FENTON PROCESS

The Fenton process, named after its developer Henry John Horstman Fenton, was first discovered in 1894. Fenton, a British chemist, discovered that when a solution of hydrogen peroxide and ferrous iron was exposed to light, it produced a powerful oxidizing agent capable of degrading organic pollutants in water. Fenton's original work focused on the use of the process to treat water contaminated with organic compounds, such as dye waste from textile mills.

The Fenton process remained largely forgotten until the 1930s when researchers began to investigate the potential of the process for water treatment. However, it was not until the 1980s that the Fenton process began to gain wider recognition as a viable method for wastewater treatment.

Since then, numerous studies have been conducted to investigate the effectiveness of the Fenton process for the treatment of various types of wastewater, including industrial and municipal wastewater. The process is effective in the degradation of a wide range of organic pollutants, including phenols, pesticides, dyes, and pharmaceuticals.

In recent years, there has been increasing interest in the use of the Fenton process with iron nanoparticles, which offers several advantages over traditional Fenton chemistry. Iron nanoparticles have a larger surface area than traditional iron catalysts, which can enhance the efficiency of the process. Additionally, the use of iron nanoparticles can reduce the amount of iron required for the process, potentially reducing costs. Overall, the Fenton process with iron has a long and fascinating history and continues to be an important tool for the treatment of wastewater in a variety of industrial and municipal applications.

PERSPECTIVE TO USE THE FENTON PROCESS IN THE INDUSTRY

The Fenton process is a chemical process that is widely used in wastewater treatment to remove contaminants from wastewater. The process involves the use of hydrogen peroxide and iron catalysts to produce highly reactive hydroxyl radicals, which then react with and break down organic compounds in the wastewater.

In the Fenton process, the iron catalyst plays a crucial role in initiating the reaction by reacting with hydrogen peroxide to produce highly reactive hydroxyl radicals. The hydroxyl radicals are highly effective in breaking down organic compounds in the wastewater, which makes the Fenton process highly efficient in removing organic contaminants from wastewater.

One of the advantages of using the Fenton process with iron in wastewater treatment is that it is a relatively simple and cost-effective method compared to other advanced wastewater treatment processes. The process does not require sophisticated equipment or high energy inputs, and the iron catalysts used in the process are relatively cheap and readily available.

Furthermore, the Fenton process with iron can be highly effective in removing a wide range of contaminants from wastewater, including organic compounds, dyes, pesticides, and pharmaceuticals. The process can also be easily adapted to different wastewater treatment scenarios and can be used in combination with other treatment processes for even better results.

However, it is important to note that the Fenton process with iron also has some limitations. For example, the process can be highly sensitive to pH and temperature conditions, which can affect the efficiency of the reaction. Additionally, the iron catalyst used in the process can also generate secondary pollutants, such as iron sludge, which may require additional treatment. In summary, the Fenton process with iron is a highly effective and cost-efficient method for removing organic contaminants from wastewater. While the process has some limitations, it can be a valuable tool in the wastewater treatment industry, especially when combined with other treatment processes to achieve optimal results.

- The Fenton process with iron can be used to treat different types of wastewater, including industrial, municipal, and agricultural wastewater. It has been used to treat wastewater from various industries, such as the textile, pharmaceutical, and food industries.
- The process is highly efficient in removing organic pollutants, which are often difficult to treat using conventional methods. Studies have shown that the Fenton process can remove up to 90% of organic pollutants from wastewater.
- The Fenton process with iron can be optimized by adjusting the dosage of hydrogen peroxide, iron catalysts, and other operating conditions such as pH, temperature, and reaction time. This can help to improve the efficiency of the process and reduce the generation of secondary pollutants.
- In addition to removing organic pollutants, the Fenton process with iron can also be used to remove heavy metals from wastewater. The process can oxidize and precipitate heavy metals, such as copper, lead, and zinc, making them easier to remove from the wastewater.
- One of the challenges of using the Fenton process with iron is the generation of iron sludge, which can be difficult to handle and dispose of. However, there are methods for treating and recycling iron sludge, which can help to reduce the environmental impact of the process.

- The Fenton process with iron can be combined with other treatment processes, such as activated sludge, membrane filtration, or adsorption, to achieve better results. This can help to remove different types of contaminants and achieve a higher level of wastewater treatment.

Overall, the Fenton process with iron is a promising method for treating wastewater and removing organic and inorganic pollutants. It offers several advantages, such as low cost, simplicity, and versatility, which make it an attractive option for wastewater treatment plants and industries. However, it is important to consider the limitations and challenges of the process and optimize it to achieve optimal results.

In summary, the Fenton process with iron offers several potential perspectives for industries seeking a cost-effective and efficient method for treating their wastewater. By optimizing the process and integrating it with other treatment methods, industries can improve their environmental performance and meet regulatory requirements.

Why Fenton process could be useful in the industry?

- Cost-effectiveness: The Fenton process with iron is relatively cost-effective compared to other advanced treatment processes, such as membrane filtration or reverse osmosis. This makes it an attractive option for industries that produce large volumes of wastewater and are looking for a cost-effective treatment method.
- Versatility: The Fenton process with iron can be used to treat wastewater from different industries, including textile, pharmaceutical, food and beverage, and chemical industries. This versatility makes it a valuable tool for industries that produce diverse types of wastewater.
- Efficiency: The Fenton process with iron is highly efficient in removing organic and inorganic pollutants from wastewater. This makes it a valuable option for industries that produce wastewater with high levels of pollutants, such as organic dyes, pharmaceuticals, or heavy metals.
- Sustainability: The Fenton process with iron can be combined with other treatment processes, such as membrane filtration or adsorption, to achieve a higher level of treatment efficiency. This can help industries meet environmental regulations and reduce the environmental impact of their wastewater discharge.
- Innovation: The Fenton process with iron is a promising area of research and development for improving the efficiency and sustainability of wastewater treatment. Researchers are exploring ways to optimize the process by using different types of catalysts or by developing new methods for treating iron sludge generated by the process.
- Public perception: Industries that implement sustainable and environmentally-friendly practices, such as the Fenton process with iron, can improve their public image and enhance their reputation as socially responsible businesses.
- Scalability: The Fenton process with iron can be easily scaled up or down, making it a flexible option for industries of different sizes. Small-scale industries can adopt the process for their wastewater treatment needs,

while large-scale industries can implement it as part of their wastewater treatment plant.

- Process integration: The Fenton process with iron can be integrated with existing wastewater treatment processes, such as activated sludge or biological treatment, to improve the overall treatment efficiency. This can help industries save on capital costs and reduce the environmental footprint of their wastewater treatment.
- Water reuse: The Fenton process with iron can help industries treat their wastewater to a level suitable for water reuse. Treated wastewater can be reused for non-potable purposes such as irrigation, cooling, or fire protection. This can help industries reduce their water consumption and minimize their impact on water resources.
- Regulatory compliance: The Fenton process with iron can help industries comply with local and national regulations related to wastewater discharge. By treating their wastewater to a level that meets regulatory standards, industries can avoid fines and legal penalties.
- Research and development: The Fenton process with iron is an area of active research and development, with scientists exploring new catalysts, reaction conditions, and optimization strategies. Industries that invest in research and development of the process can gain a competitive advantage and stay ahead of regulatory requirements.
- Resource recovery: The Fenton process with iron can be used for resource recovery, such as recovering energy or nutrients from wastewater. This can help industries reduce their dependence on external resources and improve their sustainability.

PERSPECTIVE OF THE FENTON PROCESS IN SCIENCE

The Fenton process with iron has gained a lot of attention in the scientific community due to its potential in the field of environmental engineering. Here are some perspectives on the Fenton process with iron in the scientific community:

- Novel catalyst development: Scientists are continuously exploring new catalysts that can be used in the Fenton process with iron to improve its efficiency and selectivity in treating various pollutants in wastewater. For example, recent studies have investigated the use of modified iron-based catalysts, such as zero-valent iron, bimetallic catalysts, and magnetic nanoparticles, for the Fenton process.
- Mechanism and kinetics studies: The Fenton process with iron involves complex chemical reactions that are still not fully understood. Scientists are conducting studies to unravel the mechanism and kinetics of the Fenton process, which can lead to better optimization and control of the process. The development of kinetic models can help predict the behavior of the process under different operating conditions and enable better process design.
- Optimization strategies: The Fenton process with iron can be optimized by adjusting various parameters such as pH, temperature, the concentration of

reactants, and catalyst type. Scientists are developing optimization strategies to improve the efficiency and selectivity of the process while minimizing the generation of byproducts.

- Integration with other processes: The Fenton process with iron can be integrated with other wastewater treatment processes, such as membrane filtration, adsorption, or biological treatment, to improve the overall treatment efficiency. Scientists are investigating the synergistic effects of combining the Fenton process with other processes and developing novel hybrid processes to maximize the benefits of each process.
- Environmental sustainability: The Fenton process with iron can help industries meet environmental regulations and reduce the environmental impact of their wastewater discharge. Scientists are evaluating the environmental sustainability of the Fenton process, including the potential for resource recovery, energy efficiency, and the generation of byproducts or secondary pollutants.
- Scale-up and commercialization: The Fenton process with iron has shown promising results at the laboratory scale, and scientists are working on scaling up the process for industrial applications. The commercialization of the Fenton process with iron requires further development of efficient and cost-effective catalysts, optimization strategies, and scale-up protocols.
- Reaction mechanism: The Fenton process with iron involves the generation of highly reactive hydroxyl radicals that can degrade various pollutants in wastewater. However, the detailed mechanism of hydroxyl radical formation is still not fully understood. Scientists are investigating the reaction pathways and intermediate species involved in the Fenton process, which can help improve the process efficiency and control.
- Catalyst stability: The stability of iron-based catalysts is a major concern for the Fenton process, as the catalysts can deactivate or agglomerate over time, leading to reduced activity and selectivity. Scientists are developing new catalyst synthesis methods and surface modification techniques to improve the stability and lifespan of iron-based catalysts.
- Treatment of complex matrices: The Fenton process with iron can effectively treat various pollutants in wastewater, including organic compounds, heavy metals, and pharmaceuticals. However, the treatment of complex matrices such as industrial wastewater, hospital wastewater, or agricultural runoff presents unique challenges due to the presence of diverse and high concentrations of pollutants. Scientists are developing tailored Fenton processes and hybrid processes to treat these complex matrices.
- Process monitoring and control: The Fenton process with iron requires careful monitoring and control of various parameters such as pH, temperature, and reactant concentrations to ensure efficient and selective pollutant degradation. Scientists are developing advanced sensing and control systems that can monitor and adjust the Fenton process in real time, improving its efficiency and reliability.
- Health and safety: The Fenton process with iron involves the use of reactive and potentially hazardous species such as hydroxyl radicals and

hydrogen peroxide. Scientists are investigating the health and safety risks associated with the Fenton process, including the potential for exposure to harmful byproducts or secondary pollutants.
- Application to emerging contaminants: Emerging contaminants such as microplastics, nanomaterials, and per- and polyfluoroalkyl substances (PFAS) pose a significant challenge to wastewater treatment. Scientists are exploring the application of the Fenton process with iron to treat these emerging contaminants, which can lead to the development of new and sustainable treatment technologies.

In summary, the Fenton process with iron is a promising area of research in environmental engineering, with scientists exploring new catalysts, optimization strategies, and integration with other processes. The development of the Fenton process with iron for industrial applications requires further research and development, and the scientific community is working towards optimizing the process for practical and sustainable applications.

The Perspective of the Iron Catalysts in the Fenton Process

Iron-based catalysts are a key component of the Fenton process, as they play a critical role in the generation of hydroxyl radicals that can effectively degrade pollutants in wastewater. Here are some perspectives on the use of iron catalysts in the Fenton process:

- Catalyst synthesis and modification: Iron-based catalysts can be synthesized using various methods such as co-precipitation, sol-gel, impregnation, or electrochemical deposition. Scientists are exploring new synthesis methods and modifying the surface properties of iron catalysts to improve their activity, stability, and selectivity in the Fenton process. For example, the addition of other metals or metal oxides such as Cu, Co, or Ti can enhance the catalytic activity of iron-based catalysts.
- Reactivity and selectivity: The activity and selectivity of iron-based catalysts depend on various factors such as the crystal structure, particle size, and surface area. Scientists are investigating the reactivity and selectivity of different iron catalysts under various operating conditions to optimize the Fenton process for specific pollutant types and concentrations.
- Mechanism and kinetics: The Fenton process with iron involves complex chemical reactions that are still not fully understood. Scientists are conducting studies to unravel the mechanism and kinetics of the Fenton process with iron catalysts, which can help optimize and control the process. The development of kinetic models can help predict the behavior of the process under different operating conditions and enable better process design.
- Stability and regeneration: Iron-based catalysts can deactivate or agglomerate over time, leading to reduced activity and selectivity. Scientists are developing regeneration protocols and surface modification techniques to improve the stability and life span of iron-based catalysts.

- Integration with other processes: The Fenton process with iron can be integrated with other wastewater treatment processes, such as membrane filtration, adsorption, or biological treatment, to improve the overall treatment efficiency. Scientists are investigating the synergistic effects of combining the Fenton process with other processes and developing novel hybrid processes to maximize the benefits of each process.
- Environmental sustainability: Iron-based catalysts are relatively abundant and inexpensive compared to other catalysts such as platinum or palladium. Scientists are evaluating the environmental sustainability of iron-based catalysts, including the potential for resource recovery, energy efficiency, and the generation of byproducts or secondary pollutants.
- Scale-up and commercialization: The Fenton process with iron has shown promising results in lab-scale studies, but its commercial viability requires further investigation. Scientists are conducting pilot-scale and industrial-scale studies to evaluate the technical and economic feasibility of the Fenton process with iron in large-scale wastewater treatment plants. The development of cost-effective and efficient reactor designs, catalyst preparation techniques, and process control systems are essential for the commercialization of the Fenton process with iron.
- Treatment of emerging contaminants: The Fenton process with iron has shown effectiveness in treating various pollutants in wastewater, including organic compounds, heavy metals, and pharmaceuticals. However, emerging contaminants such as microplastics, per- and polyfluoroalkyl substances (PFAS), and pharmaceutical residues require further investigation. Scientists are exploring the application of the Fenton process with iron to treat emerging contaminants and developing hybrid processes to enhance treatment efficiency.
- Application in decentralized systems: Wastewater treatment in remote and underserved communities is a critical challenge that requires innovative and cost-effective solutions. The Fenton process with iron can be adapted to decentralized wastewater treatment systems such as constructed wetlands, decentralized reactors, or on-site treatment systems. Scientists are investigating the feasibility and performance of the Fenton process with iron in decentralized systems and developing novel reactor designs and management strategies.
- Integration with renewable energy: The Fenton process with iron requires a source of energy to drive the chemical reactions. The integration of renewable energy sources such as solar, wind, or hydropower can improve the sustainability and cost-effectiveness of the Fenton process with iron. Scientists are exploring the application of renewable energy sources to power the Fenton process with iron and developing integrated systems that combine energy generation and wastewater treatment.
- Health and safety considerations: The Fenton process with iron involves the use of potentially hazardous species such as hydrogen peroxide and hydroxyl radicals. The generation and control of these species require careful consideration of health and safety risks to workers and the

environment. Scientists are investigating the potential health and safety hazards associated with the Fenton process with iron and developing guidelines and protocols to mitigate these risks.

- Public perception and acceptance: The Fenton process with iron, like any other wastewater treatment technology, requires public acceptance and support for its implementation. Scientists are communicating the benefits and limitations of the Fenton process with iron to the public, engaging stakeholders, and addressing any concerns or misconceptions. The development of sustainable and effective wastewater treatment technologies such as the Fenton process with iron can improve public health, environmental sustainability, and social equity.
- In summary, the Fenton process with iron has significant potential for wastewater treatment, and scientists are working on addressing various technical, economic, environmental, and social challenges associated with its implementation. The application of the Fenton process with iron in large-scale wastewater treatment plants, decentralized systems, and emerging contaminants requires further research and development. The integration of renewable energy sources, health and safety considerations, and public acceptance are important factors for the successful implementation of the Fenton process with iron.
- Advanced oxidation processes: The Fenton process with iron is one of the advanced oxidation processes (AOPs) that involve the production of highly reactive oxidizing species to degrade pollutants in wastewater. Scientists are exploring the synergistic effects of combining the Fenton process with iron with other AOPs such as photocatalysis, ozonation, and electrochemical oxidation. These hybrid processes can enhance the treatment efficiency, reduce the formation of toxic by-products, and improve the energy efficiency of the wastewater treatment process.
- Nanotechnology: Nanotechnology has shown potential for enhancing the efficiency of the Fenton process with iron by increasing the surface area and reactivity of the catalysts. Scientists are developing novel iron-based nanoparticles with high catalytic activity, stability, and selectivity for the Fenton process. These nanoparticles can also be functionalized with other materials to enhance their performance and specificity for target pollutants.
- Artificial intelligence and machine learning: The Fenton process with iron involves complex chemical reactions and interactions that require precise control and monitoring of various parameters such as pH, temperature, and concentration of reactants. Artificial intelligence (AI) and machine learning (ML) algorithms can improve the process control, optimization, and prediction of the Fenton process with iron. Scientists are developing AI and ML models to predict the treatment efficiency and optimize the operating conditions of the Fenton process with iron.
- Resource recovery: Wastewater treatment generates various byproducts such as sludge, biofilm, and nutrient-rich effluent. These byproducts can be recovered and reused for various purposes such as energy production,

agricultural fertilizers, and bioplastics. The Fenton process with iron can generate iron-rich sludge that can be used as a catalyst in other industrial processes or for soil remediation. Scientists are exploring the potential for resource recovery from the Fenton process with iron and developing innovative techniques for byproduct recovery and reuse.

- International collaboration and standardization: The application of the Fenton process with iron for wastewater treatment is a global challenge that requires international collaboration and standardization. Scientists are working on developing international standards and guidelines for the design, operation, and monitoring of the Fenton process with iron. The collaboration between academia, industry, and government can accelerate the development and implementation of the Fenton process with iron for sustainable and efficient wastewater treatment.
- Overall, the Fenton process with iron has shown significant potential for wastewater treatment, and scientists are exploring various avenues to improve its efficiency, sustainability, and commercial viability. The application of advanced technologies such as nanotechnology, AI and ML, and hybrid processes can enhance treatment efficiency and reduce the environmental impact of wastewater treatment. The recovery of by-products and the collaboration between stakeholders can create a circular economy for wastewater treatment, reducing the dependence on finite resources and improving the sustainability of the wastewater treatment process.

In summary, the Fenton process with iron has significant potential for wastewater treatment, and scientists are working on addressing various technical, economic, environmental, and social challenges associated with its implementation. The application of the Fenton process with iron in large-scale wastewater treatment plants, decentralized systems, and emerging contaminants requires further research and development. The integration of renewable energy sources, health and safety considerations, and public acceptance are important factors for the successful implementation of the Fenton process with iron.

The Perspective of the Nanostructures Catalysts in the Fenton Process

Iron nanostructures catalysts have recently emerged as a promising candidate for the Fenton process in wastewater treatment. These catalysts have unique physical and chemical properties, such as a high surface-area-to-volume ratio, which enhances their catalytic activity and selectivity for specific pollutants.

One of the key advantages of iron nanostructured catalysts is their high reactivity towards hydrogen peroxide, which results in the formation of hydroxyl radicals ($^{\bullet}OH$). These hydroxyl radicals are highly reactive and can effectively oxidize a wide range of organic pollutants, including pharmaceuticals, dyes, and pesticides, present in wastewater. Iron nanostructures catalysts can be synthesized in various forms, including nanoparticles, nanowires, nanotubes, nanosheets, and nanorods. Each form has unique properties that can be tailored for specific wastewater treatment applications. For instance, nanowires have a high aspect ratio that enhances their transport properties, while nanotubes have a hollow structure that can be used for the selective removal of organic pollutants.

Iron nanostructured catalysts can also be functionalized with other elements, such as carbon or nitrogen, to enhance their catalytic activity and selectivity. For instance, nitrogen-doped iron nanoparticles have been shown to have enhanced catalytic activity for the degradation of organic pollutants compared to undoped iron nanoparticles.

One of the significant advantages of iron nanostructured catalysts is their ease of separation from the treated wastewater. They can be readily removed from the treated wastewater using magnetic fields, making the separation process simple and efficient. This reduces the need for additional separation steps and makes the treatment process more cost-effective. Despite the advantages of iron nanostructures catalysts in the Fenton process, there are also challenges that need to be addressed. These include the potential toxicity of the catalysts, the stability of the catalysts under harsh treatment conditions, and the scale-up of the synthesis process for commercial application. Overall, the perspective of iron nanostructures catalysts in the Fenton process is promising, and further research is needed to optimize their synthesis, performance, and scale-up for commercial application in wastewater treatment. The development of sustainable and efficient iron nanostructures catalysts for the Fenton process can significantly contribute to the advancement of wastewater treatment technologies and the protection of the environment.

Iron nanostructures catalysts have shown promising results for the Fenton process in various wastewater treatment applications, including the degradation of organic pollutants and the removal of heavy metals. In addition to their high reactivity and selectivity, iron nanostructures catalysts have shown excellent stability under harsh treatment conditions. They can also be synthesized using simple and cost-effective methods, such as chemical reduction and hydrothermal synthesis. Iron nanostructures catalysts can also be combined with other advanced oxidation processes, such as photocatalysis, to enhance the degradation of organic pollutants. For instance, the combination of iron oxide nanoparticles and titanium dioxide nanoparticles has been shown to improve the degradation of pollutants under visible light irradiation.

Another advantage of iron nanostructures catalysts is their ability to reduce the formation of toxic by-products during the Fenton process. This is because the hydroxyl radicals generated by the iron nanostructures catalysts are highly reactive and can efficiently oxidize the organic pollutants into simpler and less toxic compounds.

Furthermore, iron nanostructures catalysts can also be used in the treatment of wastewater containing heavy metals. Iron nanostructures catalysts can effectively remove heavy metals from wastewater through adsorption, precipitation, and redox reactions. This is due to the strong affinity of heavy metals towards iron species and the ability of iron nanostructures catalysts to generate hydroxyl radicals that can oxidize the heavy metal ions into less toxic forms.

PERSPECTIVE OF THE ECONOMIC ASPECT IN FENTON PROCESS

The Fenton process with iron has gained significant attention in recent years due to its potential for cost-effective treatment of wastewater contaminated with organic pollutants. The economic perspective of the Fenton process with iron in wastewater treatment considers the costs associated with the process, including the cost of chemicals, equipment, and labor, and the potential economic benefits of the process.

One of the significant advantages of the Fenton process with iron is that it is a relatively simple and low-cost technology compared to other advanced oxidation processes. The process requires only two primary chemicals, hydrogen peroxide and iron, and does not require the use of expensive catalysts or UV lamps.

Furthermore, the Fenton process with iron can be operated at ambient temperature and pressure, which significantly reduces the energy costs associated with heating and pressurizing the wastewater. The process can also be easily integrated into existing wastewater treatment plants, minimizing the need for additional infrastructure and equipment. The use of iron as a catalyst in the Fenton process is also cost-effective compared to other catalysts such as copper, manganese, and cobalt. Iron is readily available and less expensive than other metals, making it an attractive option for large-scale wastewater treatment applications.

Another economic advantage of the Fenton process with iron is its high efficiency in the removal of organic pollutants. The process can achieve high removal rates of organic pollutants, reducing the need for additional treatment steps and minimizing the costs associated with the disposal of residual pollutants. In addition to the cost savings associated with the treatment process itself, the Fenton process with iron can also generate economic benefits through the recovery of valuable by-products. Iron sludge produced during the process can be recycled and reused as a coagulant in wastewater treatment, reducing the need for additional chemicals and minimizing disposal costs.

Overall, the economic perspective of the Fenton process with iron in wastewater treatment is promising, with significant cost savings and potential economic benefits. However, it is essential to consider the scalability of the process and the potential environmental impacts of the iron sludge generated during the treatment process. Further research is needed to optimize the process and develop sustainable and cost-effective solutions for large-scale wastewater treatment applications.

Another economic advantage of the Fenton process with iron is its versatility and adaptability to a wide range of wastewater treatment applications. The process can be used to treat various types of wastewater, including industrial effluents, municipal wastewater, and agricultural wastewater.

Moreover, the Fenton process with iron can be easily scaled up or down depending on the wastewater treatment requirements. This flexibility makes it an attractive option for small and medium-sized enterprises that may not have the resources to invest in expensive treatment technologies. In terms of operational costs, the Fenton process with iron can also offer economic benefits by reducing the need for additional treatment steps or chemicals. For example, the process can effectively remove organic pollutants and pathogens from wastewater, reducing the need for additional disinfection steps or chemical dosing. Additionally, the Fenton process with iron can also reduce the environmental impact of wastewater treatment, which can result in long-term economic benefits. By effectively removing organic pollutants, the process can minimize the environmental impact of untreated wastewater on ecosystems and reduce the risk of waterborne diseases.

Finally, the Fenton process with iron can also generate revenue from carbon credits. The process can be considered a carbon-neutral technology since the hydroxyl radicals generated during the process can oxidize organic pollutants into carbon dioxide and water. The reduction in greenhouse gas emissions can be monetized through carbon credits, providing additional economic benefits for wastewater treatment facilities.

In summary, the economic perspective of the Fenton process with iron in wastewater treatment is promising, with significant cost savings, operational efficiencies, and potential revenue streams. However, it is crucial to balance the economic benefits of the process with its potential environmental impacts and ensure sustainable and responsible wastewater treatment practices.

CHALLENGES IN THE FENTON PROCESS

Although the Fenton process with iron has proven to be a promising technology for wastewater treatment, it still faces several challenges that need to be addressed to ensure its successful implementation in industrial settings.

- One of the main challenges is the optimization of the process conditions, particularly the selection of the appropriate dosage and pH of the reactants. The optimal dosage and pH will depend on several factors, such as the type and concentration of the contaminants, the type of iron catalyst used, and the characteristics of the wastewater. Achieving optimal process conditions can significantly improve the efficiency of the process and reduce treatment costs.
- Another challenge is the formation of sludge during the process, which can pose environmental and economic issues. The sludge contains the treated pollutants and needs to be properly managed and disposed of to avoid negative impacts on the environment. There are several approaches to managing the sludge, such as chemical stabilization, composting, and landfilling, but each method has its limitations, and the selection of the appropriate method will depend on the specific wastewater characteristics and local regulations.
- The Fenton process with iron also requires a continuous supply of hydrogen peroxide, which can be expensive and may pose safety risks if not handled properly. Finding a cost-effective and safe source of hydrogen peroxide is critical for the successful implementation of the Fenton process with iron in industrial settings.
- Furthermore, the efficiency of the Fenton process with iron can be impacted by the presence of other contaminants, such as salts and heavy metals, in the wastewater. These contaminants can interfere with the reaction between the hydrogen peroxide and the iron catalyst, resulting in reduced efficiency and increased treatment costs.
- Another challenge in the Fenton process with iron is the potential formation of toxic byproducts, such as hydroxylamine and nitrous oxide, which can pose health and environmental risks. The formation of these by-

products can be influenced by several factors, such as the pH and temperature of the reaction, the concentration of the iron catalyst, and the presence of other contaminants in the wastewater. The identification and control of these by-products are crucial for the safe and effective implementation of the Fenton process with iron.

- In addition, the scale-up of the Fenton process with iron from laboratory to industrial scale can be challenging, as the process conditions that work well at the laboratory scale may not be effective at a larger scale. For instance, the homogeneity and mixing of the reactants become critical at larger scales to ensure the proper contact between the iron catalyst and the pollutants. Therefore, optimizing the process parameters for large-scale applications is necessary to ensure the efficiency and effectiveness of the Fenton process with iron.
- Finally, the Fenton process with iron may not be suitable for the treatment of all types of wastewater, as some contaminants may be resistant to the process or require different treatment methods. Thus, a comprehensive understanding of the characteristics of the wastewater and the nature of the contaminants is crucial to determine the suitability of the Fenton process with iron and to identify the appropriate treatment methods.
- Despite these challenges, the Fenton process with iron remains a promising technology for wastewater treatment due to its high efficiency, low cost, and ability to treat a wide range of contaminants. Addressing these challenges and developing appropriate strategies to overcome them will be critical to ensure the successful implementation of the Fenton process with iron in industrial wastewater treatment and to promote sustainable and effective wastewater management practices.

FENTON PROCESS IN LABORATORY

The Fenton process can be a great activity for students to engage in for several reasons. First, it offers a hands-on opportunity for students to learn about important concepts in chemistry and environmental science. Students can gain a deeper understanding of chemical reactions, oxidation-reduction, and the principles of wastewater treatment by conducting experiments with the Fenton process. Additionally, the Fenton process can be a valuable tool for teaching students about the importance of sustainability and environmental stewardship. By demonstrating the potential of the Fenton process to treat wastewater effectively and sustainably, students can gain a better understanding of how science and technology can be used to address environmental challenges.

The Fenton process can also be a great way to teach students about the importance of interdisciplinary thinking and problem-solving. The Fenton process involves concepts from chemistry, environmental science, and engineering, and requires students to integrate these disciplines to design and conduct effective experiments.

Overall, the Fenton process is a valuable activity for students that offers opportunities for hands-on learning, interdisciplinary thinking, and engagement with important environmental issues. It can be a great way to inspire students to pursue careers in science and technology, and to promote sustainability and environmental stewardship.

- Introduction: Begin by introducing the Fenton process and its applications in wastewater treatment. Discuss the principles of the process, including the role of iron catalysts, hydrogen peroxide, and pH control.
- Safety precautions: Review the safety precautions that need to be taken when working with chemicals, including wearing appropriate personal protective equipment, working in a well-ventilated area, and avoiding direct contact with chemicals.
- Experiment setup: Set up the experiment by preparing the wastewater sample, adding the iron catalyst, and adding hydrogen peroxide to initiate the Fenton reaction. Monitor the pH and adjust as necessary to ensure optimal reaction conditions.
- Data collection: Collect data on the effectiveness of the Fenton process in removing contaminants from the wastewater sample. Use analytical techniques such as spectrophotometry or titration to measure the concentration of target contaminants before and after treatment.
- Analysis: Analyze the data to evaluate the effectiveness of the Fenton process in removing contaminants from the wastewater sample. Discuss the limitations of the process and potential areas for improvement.
- Conclusion: Conclude the class by discussing the broader implications of the Fenton process for wastewater treatment and environmental sustainability. Encourage students to think about how they can apply the principles they have learned to real-world environmental challenges.

To make a Fenton reaction with students, you will need:

- Hydrogen peroxide: a 30% solution of hydrogen peroxide is commonly used in Fenton reactions.
- Iron catalyst: Iron sulfate ($FeSO_4$) or iron chloride ($FeCl_3$) are commonly used catalysts in the Fenton process.
- Wastewater sample: obtain a sample of wastewater to be treated.
- pH meter or pH test strips: to monitor and adjust the pH of the reaction mixture.
- Glassware: beakers, flasks, pipettes, and stirring rods for mixing and handling the reaction mixture.
- Safety equipment: goggles, gloves, and lab coats to protect students from potential hazards.

In addition, you may need access to a laboratory with appropriate facilities and equipment, including a fume hood and water supply.

It is important to ensure that all equipment and materials are properly labeled and stored, and that students are trained in proper handling and disposal procedures for chemicals and waste.

Overall, the Fenton process can be an engaging and educational activity for students, allowing them to learn about the principles of wastewater treatment and environmental sustainability while gaining practical experience in the laboratory.

USING FENTON PROCESS IN INDUSTRY

If you are interested in applying the Fenton process with iron in your company for wastewater treatment, there are several steps you can take:

- Conduct a feasibility study: Before implementing the Fenton process, it is important to assess the suitability of the technology for your specific wastewater treatment needs. A feasibility study can help determine the potential benefits, limitations, and costs associated with the Fenton process.
- Design and engineering: Once the feasibility study has been completed, the next step is to design the Fenton process system and engineer the necessary equipment and infrastructure. This includes selecting the appropriate reactor type, determining the optimal process conditions, and designing the necessary piping and instrumentation.
- Installation and testing: After the design and engineering phase, the Fenton process system must be installed and tested to ensure proper operation and compliance with local regulations. This includes the installation of the necessary piping, instrumentation, and control systems.
- Operation and maintenance: Once the Fenton process system is operational, it is important to establish an ongoing operation and maintenance program to ensure reliable and effective performance. This includes routine monitoring of process conditions, maintenance of equipment, and regular testing of effluent quality.
- Regulatory compliance: It is important to ensure that the Fenton process system complies with local regulations and permits for wastewater treatment. This includes obtaining necessary permits, monitoring and reporting effluent quality, and maintaining compliance with regulatory standards.

WHAT DO YOU HAVE TO REMEMBER ABOUT?

- Wastewater characterization: Before implementing the Fenton process, it is important to understand the composition and characteristics of the wastewater being treated. This can help determine the appropriate process conditions and optimize the treatment efficiency.
- Iron catalyst selection: There are different types of iron catalysts that can be used in the Fenton process, including zero-valent iron, iron oxide, and iron-containing nanoparticles. The selection of the appropriate iron catalyst depends on factors such as the type of wastewater being treated, the desired treatment efficiency, and the cost.
- pH control: The Fenton process is highly pH-dependent, and the optimal pH range for the process is typically between 2.5 and 4.0. It is important to

monitor and control the pH of the wastewater being treated to ensure optimal performance of the process.

- Hydrogen peroxide dosing: The amount of hydrogen peroxide dosed into the system is a critical parameter for the Fenton process. The optimal dosage depends on factors such as the concentration of organic pollutants in the wastewater, the iron catalyst concentration, and the desired treatment efficiency.
- Safety considerations: The Fenton process involves the use of hydrogen peroxide, which can be hazardous if not handled properly. It is important to implement appropriate safety measures and protocols to ensure safe operation of the process.

FENTON PROCESS IN THE FUTURE

- Use of alternative oxidants: While hydrogen peroxide is the most commonly used oxidant in the Fenton process, researchers are exploring the use of alternative oxidants, such as persulfate and peroxymonosulfate. These oxidants have the potential to improve the efficiency and effectiveness of the process, while also reducing costs and environmental impact.
- Integration with other treatment technologies: The Fenton process can be used in combination with other treatment technologies, such as activated carbon adsorption and membrane filtration, to further enhance the removal of pollutants from wastewater. Research in this area is focused on identifying the most effective treatment combinations and optimizing the integration of different technologies.
- Use of advanced catalysts: In addition to iron catalysts, researchers are exploring the use of other advanced catalysts, such as titanium dioxide and manganese oxide, in the Fenton process. These catalysts have the potential to enhance the efficiency of the process, reduce costs, and improve the environmental sustainability of wastewater treatment.
- Application in emerging areas: The Fenton process is being investigated for use in emerging areas, such as the treatment of emerging contaminants, such as microplastics and pharmaceuticals, and the treatment of industrial wastewaters containing complex mixtures of pollutants. This research is focused on identifying the most effective treatment approaches and developing new technologies to meet these challenges.

While the Fenton process is primarily used for wastewater treatment, recent studies have investigated the potential for using the process to produce hydrogen gas as a source of renewable energy. The Fenton process involves the generation of hydroxyl radicals from the reaction between hydrogen peroxide and ferrous ions. These hydroxyl radicals are highly reactive and can break down organic contaminants in the wastewater, but they can also react with water molecules to produce hydrogen gas.

Researchers have explored different approaches to optimize the Fenton process for hydrogen production. For example, some studies have focused on enhancing the production of ferrous ions or hydrogen peroxide to increase the yield of hydroxyl radicals and hydrogen gas. Other studies have investigated the use of different types of iron catalysts or the addition of other chemicals to enhance the process efficiency.

The potential for using the Fenton process for hydrogen production has significant implications for sustainable energy generation. Hydrogen is a clean and renewable source of energy that can be used in a variety of applications, including fuel cells, combustion engines, and industrial processes. By integrating the Fenton process with hydrogen production, wastewater treatment plants could become a source of renewable energy, reducing their reliance on fossil fuels and enhancing their sustainability.

However, there are still several challenges that need to be addressed before the Fenton process can be widely used for hydrogen production. These challenges include optimizing the process parameters, developing efficient catalysts, and addressing issues related to scalability and commercialization. Nonetheless, the potential for using the Fenton process for hydrogen production highlights the versatility and promise of this technology in addressing pressing environmental and energy challenges.

SUMMARY

The Fenton process with iron is an advanced oxidation process that has gained significant attention in recent years due to its high efficiency and low cost in treating wastewater. The process involves the generation of hydroxyl radicals, which are highly reactive and can effectively oxidize a wide range of organic pollutants in wastewater.

Iron catalysts play a critical role in the Fenton process, and recent research has focused on developing iron nanostructures as efficient and cost-effective catalysts. Iron nanostructures, such as nanoparticles, nanorods, and nanotubes, have been found to have high catalytic activity and stability, making them an attractive alternative to traditional iron catalysts. The Fenton process with iron has several advantages over other advanced oxidation processes. It is relatively low cost, and the reactants, hydrogen peroxide, and iron, are readily available and inexpensive. Additionally, the process can be easily integrated into existing wastewater treatment plants, making it an attractive option for upgrading existing infrastructure.

The Fenton process is a chemical oxidation process that involves the generation of hydroxyl radicals ($^{\bullet}OH$) from the reaction between hydrogen peroxide (H_2O_2) and a catalyst, usually iron. The hydroxyl radicals are strong oxidizing agents that react with organic compounds present in wastewater, resulting in their degradation.

The Fenton process is a complex process that involves a series of reactions between the catalyst, H_2O_2, and the organic compounds present in wastewater. The reaction rate depends on various factors such as pH, temperature, H_2O_2 concentration, and the nature of the organic compounds.

Another advantage of the Fenton process with iron is its versatility in treating various types of wastewater, including industrial effluents, municipal wastewater, and agricultural wastewater. The process can effectively remove organic pollutants, pathogens, and nutrients from wastewater, making it an attractive option for water reuse and resource recovery applications.

The most important kinds of the process:

- Homogeneous Fenton process: This is the original Fenton process, which involves the reaction of hydrogen peroxide with ferrous ions in acidic conditions to produce hydroxyl radicals. This process is simple and easy to implement, but can be limited by the need for low pH conditions, which can increase the cost of neutralization and can lead to corrosion issues.
- Heterogeneous Fenton process: In this process, the catalyst is supported on a solid surface, such as activated carbon, zeolite, or silica, to improve the stability and reusability of the catalyst. This process can operate at neutral or slightly acidic pH, reducing the need for pH adjustment and reducing corrosion issues.
- Photo-Fenton process: This process combines the Fenton process with UV irradiation, which can enhance the production of hydroxyl radicals and increase the degradation rate of pollutants. This process can operate at neutral or slightly acidic pH and can be effective for treating pollutants that are resistant to degradation by traditional Fenton processes.
- Electro-Fenton process: In this process, an electric current is used to generate ferrous ions from an iron anode, which then reacts with hydrogen peroxide to produce hydroxyl radicals. This process can operate at neutral or slightly acidic pH and can be useful for treating pollutants that are not easily degraded by traditional Fenton processes.
- Photo-electro-Fenton process: This process combines the photo-Fenton process with the electro-Fenton process, using UV irradiation and an electric current to generate hydroxyl radicals. This process can be effective for treating wastewater that contains refractory pollutants and can operate at neutral or slightly acidic pH.

Moreover, there are several types of iron that can be used as catalysts in the Fenton process for wastewater treatment, including:

- Ferrous iron (Fe^{2+}): This is the most commonly used iron species in the Fenton process. It is typically added as ferrous sulfate ($FeSO_4$) or ferrous chloride ($FeCl_2$) to the wastewater.
- Zero-valent iron (Fe^0): This is metallic iron in its elemental form. It can be used as a heterogeneous catalyst in the Fenton process and has been shown to be effective in treating certain types of wastewater.
- Iron oxides (FeO_x): These include iron(II) oxide (FeO), iron(III) oxide (Fe_2O_3), and magnetite (Fe_3O_4). Iron oxides can be used as homogeneous or heterogeneous catalysts in the Fenton process.
- Iron nanoparticles (FeNPs): These are tiny particles of iron with diameters in the nanometer range. They have a high surface area and can be used as heterogeneous catalysts in the Fenton process.
- Iron chelates: These are complexes formed between iron and organic molecules called chelating agents. They can be used as homogeneous catalysts in the Fenton process.

- Magnetite (Fe_3O_4): Magnetite is a naturally occurring mineral that is commonly used in the Fenton process. It has a high reactivity and can be easily separated from wastewater using a magnetic field.
- Hematite (Fe_2O_3): Hematite is another naturally occurring iron oxide that can be used in the Fenton process. It is less reactive than magnetite but can still be effective at removing contaminants from wastewater.

The choice of iron catalyst will depend on factors such as the type of wastewater being treated, the desired treatment efficiency, and the cost and availability of the catalyst.

There are several kinds of reactors that can be used in the Fenton process for wastewater treatment. The choice of reactor depends on the specific application and the desired outcome. Here are some common types of reactors used in the Fenton process:

- Stirred Tank Reactors (STR): Stirred tank reactors are the most commonly used reactors in the Fenton process. They are simple to operate and control and can handle a wide range of flow rates and concentrations. In this reactor, the reaction mixture is stirred to ensure that the iron catalyst and hydrogen peroxide are well mixed, which enhances the reaction rate.
- Fixed-Bed Reactors (FBR): Fixed-bed reactors use a stationary bed of catalyst material that allows the wastewater to flow through it. As the water passes through the bed, the catalyst oxidizes the pollutants. This type of reactor is useful for treating wastewater with high concentrations of organic pollutants.
- Batch Reactors: Batch reactors are used when treating small volumes of wastewater. In this type of reactor, a batch of wastewater is treated at once. The Fenton reagents are added to the reactor, and the mixture is stirred until the reaction is complete.
- Fluidized-Bed Reactors (FBR): Fluidized-bed reactors use a bed of catalyst particles that is suspended in the wastewater. Air is blown through the bed, creating a fluidized state. The wastewater flows through the bed, and the catalyst oxidizes the pollutants.
- Photo-Fenton Reactors: Photo-Fenton reactors use light to enhance the Fenton process. In this reactor, a light source is used to activate the catalyst and accelerate the reaction. This type of reactor is useful for treating wastewater with low concentrations of pollutants.
- Bed reactor: In this reactor, the catalyst is packed into a column or bed, through which the wastewater flows. The high surface area of the catalyst provides ample opportunity for reaction to occur.
- Rotating drum reactor: This type of reactor uses a rotating drum or cylinder filled with the catalyst and wastewater. As the drum rotates, the wastewater is agitated, allowing for efficient mixing and reaction.
- Membrane reactor: In this type of reactor, a membrane is used to separate the catalyst from the wastewater. This allows for continuous removal of the reaction products and prevents the catalyst from fouling or deactivating.

The Fenton process with iron can also generate valuable byproducts, such as iron sludge, which can be recycled and reused. Iron sludge can be used in various applications, such as soil amendments, cement production, and pigments.

In terms of economic benefits, the Fenton process with iron has the potential for cost savings, operational efficiencies, and revenue streams from carbon credits. The process can reduce the need for additional treatment steps or chemicals, resulting in cost savings. It can also generate revenue from carbon credits since the hydroxyl radicals generated during the process can oxidize organic pollutants into carbon dioxide and water, reducing greenhouse gas emissions. However, it is essential to balance the economic benefits of the Fenton process with iron with its potential environmental impacts, such as the generation of iron sludge. Iron sludge can be a significant environmental concern, and proper management is required to ensure that it is safely and responsibly handled. Additionally, the process may generate other environmental impacts, such as the generation of secondary pollutants, and these need to be carefully monitored and mitigated.

In conclusion, the Fenton process with iron is a promising technology for wastewater treatment that offers high efficiency, low cost, and versatility. The process can generate valuable by-products and has the potential for cost savings and revenue streams from carbon credits. However, it is crucial to balance the economic benefits of the process with its potential environmental impacts and ensure sustainable and responsible wastewater treatment practices. Further research is needed to optimize the process and develop sustainable solutions for large-scale wastewater treatment applications.

One of the key factors that influence the efficiency of the Fenton process with iron is the pH of the solution. The optimal pH range for the process is typically between 2.5 and 4.0, and adjusting the pH to this range can significantly enhance the efficiency of the process. However, maintaining the optimal pH range can be challenging, and additional chemicals may be required to adjust the pH of the solution.

Another important factor that affects the efficiency of the Fenton process with iron is the concentration of the reactants, hydrogen peroxide, and iron. Increasing the concentration of the reactants can enhance the efficiency of the process, but it can also increase the cost and potential environmental impacts. Therefore, it is essential to balance the concentration of the reactants with the economic and environmental factors.

The Fenton process with iron has been applied in various industries, such as the textile, paper, and pharmaceutical industries, to treat their wastewater. In the textile industry, the process has been found to effectively remove dyes and other organic pollutants from the wastewater, resulting in significant improvements in water quality. In the paper industry, the process has been used to remove lignin and other organic compounds from the wastewater, improving the efficiency of the overall treatment process. In the pharmaceutical industry, the process has been applied to remove antibiotics and other pharmaceuticals from the wastewater, reducing the risk of environmental contamination and improving public health.

Finally, it is worth noting that the Fenton process with iron is not a standalone solution for wastewater treatment and should be integrated into a broader wastewater treatment system. Other treatment steps, such as biological treatment, may be required to remove nutrients and other pollutants from the wastewater. Additionally, proper monitoring and management of the process are essential to ensure that the wastewater is treated effectively and safely. Overall, the Fenton process with iron is a promising technology for wastewater treatment that offers high efficiency, low cost, and versatility. The process has been applied in various industries to effectively remove organic pollutants and other contaminants from wastewater, resulting in significant improvements in water quality. However, it is important to balance the economic benefits of the process with its potential environmental impacts and ensure sustainable and responsible wastewater treatment practices.

The Fenton process with iron is an environmentally friendly approach to wastewater treatment that offers several benefits over traditional treatment methods. One of the key advantages of the Fenton process is that it can effectively remove a wide range of contaminants from wastewater, including organic compounds, heavy metals, and pathogens. This makes it a versatile and effective approach to treating wastewater from a variety of industrial and municipal sources.

Another environmental benefit of the Fenton process is that it does not rely on the use of harmful chemicals or generate hazardous byproducts. Unlike traditional treatment methods that often involve the use of chlorine, ozone, or other chemicals, the Fenton process only requires the use of hydrogen peroxide and iron salts. These chemicals are readily available and do not pose significant risks to human health or the environment.

In addition to being a safe and effective treatment method, the Fenton process also offers a low carbon footprint compared to other treatment methods. The process does not require significant amounts of energy or generate large amounts of greenhouse gases, making it a more sustainable and environmentally friendly approach to wastewater treatment. Furthermore, the Fenton process with iron can also be used to recover valuable resources from wastewater, such as metals and organic compounds. This can further reduce the environmental impact of wastewater treatment by reducing the need for new resource extraction and reducing the amount of waste that needs to be disposed of. Overall, the Fenton process with iron represents a promising and environmentally friendly approach to wastewater treatment that offers several benefits over traditional treatment methods. Its versatility, low carbon footprint, and potential for resource recovery make it a valuable tool for promoting sustainability and protecting the environment.

The Fenton process with iron can be an alternative to various sewage treatment methods, depending on the specific requirements and conditions of the wastewater.

One common alternative to the Fenton process is the traditional biological treatment method, which uses microorganisms to degrade organic matter in wastewater. However, this method can be limited by factors such as temperature, pH, and nutrient availability, and may not be effective in treating some types of pollutants.

The Fenton process can be particularly useful for treating wastewater that contains refractory or toxic organic compounds, as it can effectively degrade these compounds without requiring a long residence time or specialized microorganisms. Additionally, the Fenton process can be used as a pre-treatment step to enhance the efficiency of biological treatment or other methods.

Other alternative sewage treatment methods that may be used alongside or instead of the Fenton process include physical treatment methods such as filtration or sedimentation, and chemical treatment methods such as ozonation or advanced oxidation processes. The specific choice of treatment method will depend on factors such as the nature and concentration of pollutants in the wastewater, the desired effluent quality, and the available resources and infrastructure for treatment.

Index

Printed in the United States
by Baker & Taylor Publisher Services